WANT ACCESS TO SOLID REFUTING THE INCESSANT MEDIA HYPE SURROUNDING CLIMATE CHANGE? THEN THE MYTHOLOGY OF GLOBAL WARMING IS FOR YOU!

The climate activists in charge of the National Oceanic and Atmospheric Administration (NOAA) were dumbfounded. Their own temperature data as well as satellite results had been showing that the climate had been stable or cooling since 1998. Even worse, the climate was cooling while CO_2 levels continued to rise. The recent climate was not being politically correct.

It was time to resort to the strategy that progressives always use in times of crisis: if you don't like the facts, throw them out and make up your own. Since December of 2015, the web sites of both NOAA and NASA have simply eliminated what they had been calling a 'pause' in global warming. They have taken their data behind a dark curtain for 'editing' and 'reanalysis' . . . and viola. Climate data suddenly show that the Earth's temperature has increased by 1°C in just the last few years. This falsification of government climate data has not gone unnoticed...

The Mythology of Global Warming is intended to provide the general public with a broad spectrum of scientific and factual information on the subject of Climate Change. This book debunks the incessant, emotional, and largely unsubstantiated claims made by the progressive media and climate scientists that industrial societies such as the United States are destroying our planet due to the use of fossil fuels. What causes global warming? What is a greenhouse gas? What impact do carbon dioxide emissions from fossil fuels actually have on the Earth's climate relative to naturally occurring phenomena? Is all ice on Earth really melting, and are sea levels rising at a catastrophic rate? Are all forms of extreme weather, including hurricanes, tornadoes, floods, and droughts increasing dramatically? Are polar bears and other life forms being pushed to the brink of extinction? Will all of this mayhem cease if fossil fuels are replaced by 'green' renewable energy sources? Answers to these questions clearly show that hard facts do not support *any* of the above dire predictions.

i

The science of global warming is indeed 'settled'; Global Warming is a myth.

"...Global warming proponents can't prove that man is destroying the planet due to global warming, but Dr. Bunker can prove that we are not. He packs a lot of punch in this small package. Read it, and arm yourself for the great debate."—*Phil Valentine, nationally syndicated talk show host of the Phil Valentine Show on Westwood One*

"In the past 20 years I have reviewed two dozen books dealing with Anthropomorphic Global Warming. There has not been nor ever will be a more comprehensive and understandable book on this subject which is to critical to the entire world's population."—*Jay Lehr, Ph.D. Science Director, The Heartland Institute*

"This is a scholarly work written by a true scientist, yet in a way that makes the topic still accessible to the average person interested in understanding both the science and also the politics of global warming. Highly recommended."—*Dr. Jennifer Marohasy, Senior Fellow, Australia's Institute of Public Affairs, co-author of "Climate Change: The Facts, 2014"*

"Unlike so many others, Dr. Bunker's book is so much more than a supposition wrapped up in a pretty bow of meaningless numbers. If you've been waiting for a book that gives actual facts in an easily checked form, you've found it."—*G. Dedrick Robinson Ph.D., co-author of Global Warming: Alarmists, Skeptics & Deniers.*

"A timely and well researched book not only for the thoughtful engaged reader, but also for the general public. The book is up-to-date and deals honestly with continuing controversies and uncertainties."—*Dr. Sonja A. Boehmer-Christiansen, Department of Geography, Hull University, Former Editor, Energy & Environment.*

THE MYTHOLOGY OF GLOBAL WARMING

Bruce Bunker, Ph.D.

Moonshine Cove Publishing, LLC
Abbeville, South Carolina U.S.A

FIRST MOONSHINE COVE EDITION
November 2018

ISBN: 978-1-945181-47-4
Library of Congress PCN: 2017960231
Copyright © 2018 by Bruce Bunker, Ph.D.

Cover art by Vincent Chavez, Clean & Simple Studios, Albuquerque, NM; cover and interior by Moonshine Cove Staff.

About The Author

Dr. Bruce Bunker (retired) received a Ph.D. in Inorganic Chemistry from the University of Illinois. He worked as a staff scientist and technical manager in nanotechnology, advanced materials, and environmental chemistry at both Sandia National Laboratories and Pacific Northwest National Laboratory. Management positions included: 1) Supervisor of the Electronic Ceramics Division (Sandia), 2) Associate Director of Advanced Processing in the Environmental Molecular Sciences Laboratory (PNNL), and 3) Nano-Bio Thrust Leader in the Center for Integrated Nanotechnologies (Sandia). At Sandia, he led a project to develop materials and processes for removing atmospheric carbon dioxide to mitigate global warming. He has over 100 peer-reviewed publications in journals including *Science* and *Scientific American*. Awards include: 1) the Department of Energy Office of Basic Energy Sciences Award for Outstanding Scientific Accomplishment, 2) DOE's Research and Development-100 Award, 3) Sandia National Laboratories Award for Excellence, and 4) the PNNL Outstanding Team Performance Award.

To my wife Cathy and anyone else having an open mind.

Acknowledgment

I am extremely grateful to Vincent Chavez from Clean & Simple Studios for production of the cover art as well as many of the maps and figures used in this book.

I wish to thank all of those people who provided useful feedback, corrections, and criticisms of the manuscript. My greatest thanks go to my wife Cathy, who read, commented on, and helped correct multiple drafts of the text. I would also like to thank the many friends and relatives who provided useful feedback, including my brother Paul, Jim Voigt, Bill Smith, Kevin Ewsuk, Joe and Anita Evans, and Cindi Feldwisch.

Finally, I am deeply appreciative of the help and guidance provided by Gene Robinson of Moonshine Cove Publishing, LLC. He is one of the few editors with the courage to handle a book on global warming that is not 'politically correct'.

CONTENTS

Chapter 1: The Mythology of Global Warming

"A lie told often enough becomes truth." – *Lenin*[1]

A Cautionary Tale

Back in the 1970s, over a hundred articles were written warning of immanent climate change[2]. *Time* and *Newsweek* ran multiple cover stories to document how the 'evil oil companies' and the capitalist life style in the United States were causing catastrophic damage to the Earth's climate. These articles claimed that scientists were almost unanimous in their opinion that man-made climate change would "reduce agricultural productivity for the rest of the century." "Climatologists are pessimistic that political leaders will take any positive action to compensate for the climactic change"[3] (*Newsweek*, April 28, 1975). Proposed solutions in the article included such things as outlawing the internal combustion engine. This rhetoric sounds all too familiar . . . except for one little thing: *In the 1970s, environmentalists and the media were claiming that the oil companies were destroying the planet due to **global cooling**.*

Alarmist cover stories (for example: How to Survive the Coming Ice Age (*Time*, 1977[4]) included 'facts' such as: Scientists predict that the temperature of the Earth could drop by 20°F due to man-made global cooling. Dr. Murray Mitchell of the National Oceanic and Atmospheric Administration warns: "The drop in temperature that has been experienced between 1945 and 1968 has taken us one sixth of the way to Ice Age temperatures." The narrative that the uncontrolled activities of oil companies, the free market system in general, and the United States in particular were destroying our planet due to *global cooling* was repeated so loudly and so often that it gained considerable traction within the general public. Then something truly unfortunate happened. Instead of cooling as predicted by climate change advocates, the average temperature of the Earth started to *increase*. Clearly, something had to be done to rescue the climate change agenda from utter disaster. *That 'something' involved the creation of the global warming movement.*

Enter Al Gore, Jr.

When Al Gore entered Harvard as a freshman in 1965, he had no idea that he would become the most visible standard bearer of the Global Warming movement. He was never a scientist. He received a C and a D in his natural science classes, with the D coming in the class entitled 'Man's Place in Nature.'[5] However, one science class taught by Professor Roger Revelle caught his interest. Professor Revelle came to Harvard from Scripps where he was one of the first scientists to postulate that carbon dioxide emissions from the burning of fossil fuels might contribute to a warming of the planet.[6] Later in life, when he needed to establish his credibility as a member of the climate change movement, Al Gore frequently referred to Professor Revelle as his 'beloved mentor.'

After graduating from Harvard in 1969, Gore entered the law program at Vanderbilt University. He dropped out[7] when it became clear that he was not destined to become a lawyer. However, his lack of a law degree did not hurt his career. His path lay in the realm of politics. His father, Al Gore, Sr., was a powerful senator from the state of Tennessee. Senator Gore was extremely well connected, and saw to it that his son was elected to the House of Representatives. Al Gore served in the House from 1977 to 1985. He was then elected as a Senator and served from 1985 until 1993.

Senator Gore made a name for himself by supporting the progressive movement. His defining moment as a senator came in 1990, when he authored a letter to the *New York Times* entitled "To Skeptics on Global Warming."[8] This letter signaled that he was a leading government advocate for 'green' causes. However, the outspoken opinions on the environment expressed in his 1992 book *Earth in the Balance*[9] were what brought him to the attention of the entire world. The release of the book was strategically timed to coincide with the 1992 United Nations Earth Summit held in Rio de Janeiro. At the summit, which was attended by 108 heads of state and over 10,000 reporters, Gore was treated "like a rock star." Gore's high visibility brought him to the attention of Bill Clinton, who selected Gore to be his vice president. Gore served as Vice President from 1993 to 2001.

As Vice President, Al Gore was in the strongest position seen prior to President Obama to support the global warming movement. He was able to enact policies and direct funding to ensure that the climate change agenda became a top priority of the United States government. As a spokesperson, Gore preached to the United States and the world: "There is no legitimate debate. Those who do not believe in the theory of human caused Global Warming are like . . . those who believe that the Earth is flat."[10] (Gore has wisely refused to engage in any debates on global warming, legitimate or otherwise. He certainly did not debate his former 'beloved mentor' Professor Revelle. Professor Revelle announced that he no longer supported his own theory of global warming (Chapter 9) due to satellite data showing that global temperatures are not currently rising (Chapter 6). Gore's response was to call his former 'beloved mentor' a senile old man.[11])

In one of his first pronouncements as Vice President, Gore stated that the federal government "should not debate the science of global warming, but should instead focus on the implementation of national and local greenhouse gas reduction policies and activities."[12] He proceeded to back up those words by pushing an aggressive climate change agenda during the remainder of his term of office. A cornerstone of Gore's strategy was to insure that all high-ranking officials in any government agency having any intersection with policies or funding relating to climate change were in line with his vision. These agencies include: The Department of Energy (DOE), the Environmental Protection Agency (EPA), the National Science Foundation (NSF), the Department of Education, the National Oceanic and Atmospheric Administration (NOAA), and the National Aeronautics and Space Administration (NASA). For example, when physicist William Happer, Director of Energy Research at the DOE, testified before Congress in 1993 that scientific data simply did not support the hypothesis of man-made global warming, Gore saw to it that Happer was immediately fired.[13]

Vice President Gore ran for the President of the United States in 2000. He lost. No one would have guessed that Al Gore had yet to achieve his greatest accomplishments in terms of his impact for the climate change movement or his own personal fortunes.

An Inconvenient Truth

Al Gore's crowning achievements for the climate change movement came with the publication of his book *An Inconvenient Truth*[14] in 2006 followed by a movie of the same title in 2007. The movie showed such a stark vision of the apocalyptic future awaiting the Earth due to man-made global warming that it received an Oscar for the best documentary of 2007. Due to its compelling environmental message, it garnered a Nobel Peace Prize for Al Gore in 2007. Al Gore was also able to leverage his high visibility and awards into a personal fortune. When he ended his tenure as Vice President in 2001, his net worth was $2 million. By 2013, his net worth exceeded $300 million.[15]

The movie *An Inconvenient Truth* is now almost required viewing for school children. It is the equivalent to the Bible when it comes to encapsulating the global warming movement. What is it about *An Inconvenient Truth* that captured the attention of the world, and why is it deemed to be so important?

To begin with, the movie highlights two scientific facts that are irrefutable. The first fact is that the burning of fossil fuels produces carbon dioxide (CO_2), which is one of the so-called greenhouse gases (see Chapter 2) via reactions such as:

$$C \text{ (coal)} + O_2 \text{ (oxygen)} \rightarrow CO_2 \qquad (1.1)$$

$$2 \ C_8H_{18} \text{ (octane)} + 25 \ O_2 \rightarrow 16 \ CO_2 + 18 \ H_2O \text{ (water)} \qquad (1.2)$$

The carbon dioxide gas produced by our power plants and our cars enters the atmosphere. The global warming hypothesis is that greenhouse gases such as CO_2 trap heat that is trying to radiate from the Earth's surface back into outer space, thus causing the planet to warm (see Chapter 2).

The second fact is that the burning of fossil fuels is increasing at an exponential rate[16] (Fig. 1.1). In 1940, fossil fuel combustion was generating roughly one billion tons (1 giga or Gton; see Appendix 1) of CO_2 each year. (A gigaton is 200 times the weight of the Great Pyramid at Giza, which is the world's largest man-made structure.)   By the time *An Inconvenient Truth* was produced, emissions had increased to roughly 9 Gton/year. Today, the grand total stands at 33 Gton/year.

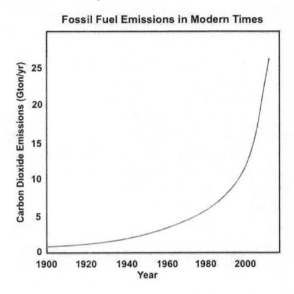

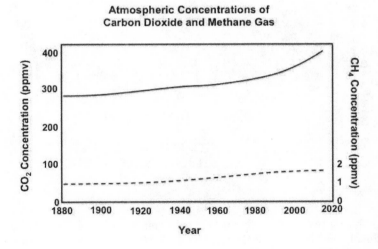

Fig. 1.1. Top - Increasing worldwide carbon dioxide emissions as a function of time during the Industrial Age (in billions of tons per year). (Data from Ref. 16) Bottom – Increasing atmospheric concentrations of carbon dioxide (solid) and methane gas (dashed)(Data from Ref. 17).

Atmospheric concentrations of greenhouse gases such as carbon dioxide and methane are also rising.[17](Fig. 1.1). These rises have been attributed to human activities such as the burning of fossil fuels and the raising of livestock. As atmospheric concentrations of greenhouse gases have been postulated to control the Earth's temperature (see Chapter 2), this means that humanity should be causing the climate to warm at an exponential rate.

Al Gore was ecstatic to discover that in 1998, Professor Michael Mann published a scientific paper in the journal *Nature* claiming to have evidence that an exponential increase in the Earth's temperature was indeed taking place.[18] In 2001, The United Nations International Panel on Climate Change (IPCC) adapted a graph from Mann's paper[19] (Fig. 1.2) to prove to the world that man-made global warming is a reality.

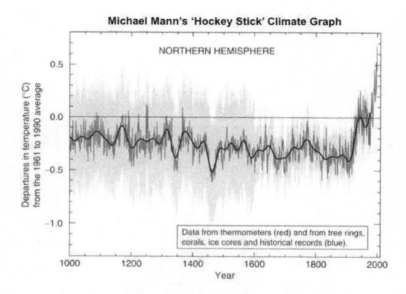

Fig. 1.2. The adaptation of Professor Michael Mann's curve currently used by the International Panel on Climate Change (IPCC) to show changes in worldwide temperatures[19] since the year 1000. Temperatures are indicated in °C. (To convert to °F, multiply by 9/5.) This famous 'hockey stick' curve suggests that the Earth's climate was essentially constant prior to the Industrial Age.

Al Gore eagerly incorporated Mann's graph into his movie, making the 'hockey stick' one of the most famous graphs of all time. When viewed side-by-side with the graph of fossil fuel emissions, it was easy for Gore and the rest of the climate change community to claim that the science of Global Warming was now 'settled.'

After showing Mann's proof that fossil fuel emissions control the Earth's temperature, *An Inconvenient Truth* moves on to provide a series of graphic images showing the apocalyptic consequences that have been predicted if man-made global warming based on the combustion of fossil fuels continues. Images include the melting of glaciers and all ice on Earth, dying polar bears, the inundation of coastal cities by massive floods, visions of entire cities being wiped off the map by more numerous and powerful hurricanes and tornados, and all food supplies being destroyed by pervasive droughts. After watching this compelling piece of climate change propaganda, it is no wonder that an entire generation has come to fear what the future might hold and to feel guilty about the role that they and their country might be having in destroying our beautiful planet.

Key Elements of the Global Warming Movement

While not every dimension of the Climate Change movement is encapsulated in *An Inconvenient Truth,* the movie sets the stage for the key elements that underpin the Global Warming narrative. These elements include:

- Carbon dioxide and methane produced from fossil fuel and the digestive systems of cows are the dominant greenhouse gases on Earth. Industrialized societies such as the United States are responsible for the production of both gases and are thus guilty of destroying our planet.
- The Earth's temperature was essentially constant until the combustion of fossil fuels and the generation of greenhouse gases in the Industrial Age caused the Earth's temperature to spiral upward and out of control.
- There is a direct correlation between atmospheric CO_2 concentrations and the temperature of the Earth. Atmospheric

carbon dioxide concentrations have never been higher than they are today due to the combustion of fossil fuels. Unless fossil fuel combustion is stopped, the Earth will become a burned out cinder.

- Global warming is causing all of the ice on Earth to melt, destroying the planet's ecosystems and inducing catastrophic rises in sea levels that are destroying coastal cities and civilizations.

- Global warming is powering new, destructive weather patterns across the globe, including the spawning of more frequent and violent hurricanes and tornados, droughts, floods, windstorms, and even record cold and snowstorms.

- Global warming is causing mass extinctions equivalent to the worst natural disasters experienced in the history of life on Earth.

- Over 95% of all scientists agree that the science of global warming is settled and that man-made climate change represents reality. Anyone who does not believe what the media and climate scientists are saying about global warming is either stupid or is a criminal who should be put in jail.

- Renewable energy sources managed by the government represent the only hope for saving the planet from global warming.

- Advocates for global warming have only one noble objective in mind, which involves protecting the Earth from evil human activities such as the burning of fossil fuels.

The above narrative should come as no surprise. Every single day people are deluged with propaganda from every possible media outlet to drive home the point that man-made global warming is destroying the Earth. Every single day is the hottest day on record. Every violent storm that is experienced is man-made. Any weather that is unusual, destructive, or unpleasant is the result of global warming. Climate change is killing any and all forms of life on the planet. Americans are constantly being told that they should feel guilty for driving their cars or using any energy to heat their homes or power their businesses. Children are depressed about the future as a result of the constant barrage of global warming warnings that they receive in school. *An entire generation has become so brainwashed on the subject that they no longer question any statement made regarding climate change.* For example, when President Obama proclaimed that there would be no terrorism in the world if it were not for global warming,[20] some Americans found themselves nodding their heads in numb agreement.

The Mythology of Global Warming

Unfortunately, essentially everything in *An Inconvenient Truth* and the climate change agenda is either a lie or a gross distortion of the truth. *The media and climate scientists have been using every means at their disposal to create what has become the Mythology of Global Warming.* When it comes to providing facts, they only provide items that support their agenda while suppressing all other pertinent information. For example, you always hear about record high temperatures, while record low temperatures are universally ignored. The media also relies on the fact that people tend to have short-term memories. For example, they know that if they claim that every single day represents a new record high temperature, few people will take the time to go back and add up all of the claims to see that the total net temperature increase is ridiculous. The media realizes that few people remember (if they ever knew) that the press was warning in the 1970's that man-made global cooling was about to plunge the earth into another Ice Age. Unfortunately, we are now in an era of 'fake news' in which media outlets can make up any story that they please knowing that: 1) few people will check their facts, and 2) there are no consequences if they are caught lying.

Fake news is one thing. Fake science is a more serious problem. It used to be that scientists performed experiments and collected data to test a scientific hypothesis. The data either supported the hypothesis, or it didn't. If the data did not support the hypothesis, scientists developed new ideas to explain all known facts. This is how science advances. However, the scientific method requires objectivity in order to progress. Unfortunately, within the climate change community, it is heresy to question the hypothesis. If the scientific data do not support global warming, the data are either thrown out or altered until the hypothesis is vindicated.

The extent to which climate data has been manipulated in recent years is both disturbing and frightening. The prime example of what has come to be known as pathological science is provided by Professor Michael Mann's graph (Fig. 1.2) that has been used for years to prove the validity of the global warming movement. Prior

9

to Michael Mann, the International Panel on Climate Change (IPCC) published a completely different, more accurate, and widely accepted graph showing how the Earth's climate has actually changed during the past thousand years[21] (Fig. 1.3). Prior to the year 1000, the Earth experienced a cool period. This was followed by an extended period from 1000 to 1400 of temperatures that were substantially higher than they are today. This Medieval Warm Period was followed by an era of colder temperatures from 1400 to around 1880 called the Little Ice Age. Although temperatures have fluctuated since the Little Ice Age, the general trend has been another period of warmer climates since 1880.

Climate Data Reported by the International Panel on Climate Change Prior to Michael Mann

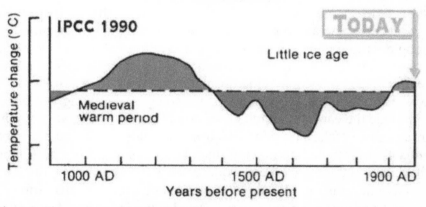

Fig. 1.3. The curve used by the IPCC as recently as 1990 to depict climate changes[21]. The data is a compilation of results from 750 scientists from over 400 research institutions in 40 countries. The results show that there have been major warming and cooling periods such as the Medieval Warm period and the Little Ice Age prior to the current fossil fuel era.

Where is the Medieval Warm Period in Michael Mann's data? Where is the Little Ice Age? Why is the only significant climate change shown associated with dramatic heating starting in 1900 at the dawn of the Industrial Age? Many scientists have since challenged and discredited Michael Mann's 'hockey stick' curve (see Chapter 9). However the most disturbing part of the 'hockey stick' story was unearthed during the so-called 'Climategate' scandal[22] involving the unauthorized release of a treasure-trove of emails between Professor Mann and other elite climate scientists (see Chapter 9). These emails show that the 'hockey stick' is a complete fabrication, involving not only the elimination of all climate data that didn't support global warming, but the inclusion of totally fictional data and data trends. This pattern of fraudulent behavior has since been mirrored by climate scientists who have been doctoring climate data entrusted to major government agencies around the world. In the United States, it has been shown that files of temperature data in the keeping of both NOAA and NASA have been manipulated to throw out results that do not agree with global warming while replacing the deleted records with computer-generated climate data (see Chapter 9). As hard as it is to believe, you can no longer trust any media report that starts with the phrase "government climate scientists say . . ." because many of these scientists can no longer be relied upon to tell the truth.

Dispelling the Myths

The objectives of the remaining chapters are to examine and dispel each of the individual myths outlined above as elements of the global warming agenda. This book provides actual data rather than unsubstantiated claims regarding climate. Data include the Earth's geologic records, historical records, meteorological data, and simple concepts of astronomy, biology, physics, and chemistry. You have probably not been exposed to most of this information, as the media and many climate scientists are not eager to share relevant climate facts with the general public. What has the Earth's temperature actually done? What influence does atmospheric carbon dioxide really have on the Earth's temperature? What influence do humans actually have on violent storms? A synopsis of key points to be made in the remaining chapters include:

Chapter 2: The dominant greenhouse gas on Earth is not the carbon dioxide from fossil fuels or methane gas emitted by cows, but water vapor. Water currently absorbs 135 times more of the Earth's outgoing heat than CO_2, and 28,000 times more heat than methane. Both CO_2 and methane are present at such low concentrations that they have a negligible impact on our climate.

Chapter 3: The temperature of the Earth has never been constant. Variations in the Earth's temperature are controlled by the amount of heat reaching us from the Sun and on how that heat is distributed. Variations in the Earth's temperature in the recent past and in the foreseeable future are both natural and predictable based on variations in solar and orbital cycles. These variations dwarf any effects that humanity has on climate.

Chapter 4: The geologic record clearly shows that there is absolutely no correlation between the Earth's temperature and atmospheric CO_2 levels. Current CO_2 levels are almost as low as they have been during the entire history of the Earth. The worst Ice Age in our planet's history occurred when CO_2 levels were twenty times higher than they are today. On the vast scale of the Earth, fossil fuel emissions are less significant than natural phenomena that add and subtract CO_2 from the atmosphere.

Chapter 5: The advance and retreat of glaciers is a natural recurring phenomenon. We are currently experiencing the high-temperature end of the latest of the 24th modern major Ice Age cycles. The massive reserves of ice on Earth have been remarkably stable in modern times. The modest retreat in glaciers observed since the end of the Little Ice Age in 1880 is normal, natural behavior. Several more advances and retreats will occur between now and the start of the next major Ice Age. Sea levels have also been remarkably stable, increasing by only 8 inches over the past 125 years. Humanity has not been threatened by the slow, incremental increases and decreases in sea level that have occurred over the course of human history.

Chapter 6: Weather patterns on Earth have been remarkably similar since the beginning of the Industrial Age. Although there are always periodic fluctuations associated with solar activity, it is not hotter, colder, wetter, or drier than it was before the combustion of large quantities of fossil fuels. Satellite measurements reveal that global temperatures have been constant since 1998. If anything, the incidence and intensity of both hurricanes and tornados has decreased in modern times. The cherry picking of individual extreme weather events by the media is just another example of dishonest reporting in support of the climate change agenda.

Chapter 7: Carbon dioxide is called a greenhouse gas for a reason. All plant life is based on a photosynthetic process involving carbon dioxide. The Earth's major food chains all depend on plant life. The geologic record clearly shows that the abundance and diversity of life has always been greater during periods in Earth's history when carbon dioxide levels and temperatures were substantially higher than they are today. Polar bears are not in danger of extinction, as the bear population has almost tripled since the 1960s.

Chapter 8: Renewable energy sources including solar and wind power are incapable of meeting the world's energy needs either now or in the future because: 1) Neither technology can physically harvest a dominant fraction of the energy that the world now uses. 2) Neither technology is cost competitive. Few countries, let alone individuals, will ever be able to afford to rely exclusively on either technology. *If renewable energy could really solve the world's energy needs, fossil fuels would have been abandoned long ago.*

Chapter 9: Global warming represents the most pervasive and damaging example of scientific fraud in world history. Progressives and the media perpetuate this fraud using a combination of misinformation and the suppression and falsification of climate data. Scientists in disagreement with global warming risk losing their ability to publish papers, receive government funding, or even stay employed. Even so, a broad spectrum of scientists and meteorologists have spoken out to say that there is no compelling evidence to support the hypothesis that humans are the cause of any climate changes observed in modern times.

Chapter 10: In reality, the global warming movement has nothing to do with 'saving the planet.' It has everything to do with the redistribution of wealth and political agendas aimed at destroying the foundations of western democracies and free market economies as exemplified by the United States of America. For this reason, it is critical for everyone to become informed on the topic of global warming. There needs to be free and open debate on this topic rather than the suppression and falsification of actual scientific climate data.

The purpose of this book is to expose you to facts that you will never hear by listening to most climate change scientists or the media. Whether you agree with these facts or not, the goal is to challenge you to think about global warming in a critical fashion rather than blindly accepting the propaganda that you are barraged with every single day. Use this information to do your own research on the topic. However, beware that most media outlets, including many web sites, Google, Facebook, Wikipedia, and the major news networks have a strong bias toward supporting the global warming movement. However, if you dig deep enough, you can check and validate all of the facts presented here.

Why Should You Care?

Global warming is not damaging your world. The global warming movement is.

The long-term goal of the movement is to unite the world under a single socialistic government in which there is no capitalism, no democracy, and no freedom. Personal freedom is fueled by access to affordable energy. As democracies and the free-market system both rely on affordable energy, a shorter-term goal is to limit the amount of energy that is available and to bring all energy sources under total government control. As fossil fuels currently power the free world, fossil fuels are in the crosshairs of the movement.

Why should you care who controls energy? Energy impacts every facet of your daily life. Whoever controls energy controls you. Below are just a few of the negative consequences that the global warming movement has already had on your life and the lives of your fellow citizens.

- *Energy costs are increasing.* Public utilities are being forced to replace existing power plants with solar and wind farms that increase the real costs of energy by as much as a factor of 7 (Chapter 8). Some of these costs are added to escalating electricity bills, which have increased by almost a factor of two since 2000. Low-income families and businesses are the biggest losers.

- *Taxes are increasing.* Taxes come out of every American's pockets. You are paying for the added costs of renewable energy whether you know it or not. Renewable energy is heavily subsidized by the government, leading to increases in federal, state, and local taxes amounting to hundreds of billions of dollars per year. Gas taxes can exceed 25%. You are paying an average of $16,000 per home to subsidize new solar installations. You paid $535 million in government subsidies for the bankruptcy of a single failed solar company Solyndra (see Chapter 8).

- *Government regulations are increasing.* For example, in just one year (2014), almost all of the 75,000 new federal regulations enacted by President Obama's administration were directed at attacking any and all companies or energy suppliers who deal with fossil fuels.[23] It is estimated that these regulations cost our economy hundreds of billions of dollars per year.

- *The objectivity of major government agencies is decreasing.* All agencies that fund or regulate any activities relating to climate, including NOAA, NASA, DOE, NSF, and the EPA have become so heavily politicized that any project, scientist, or activity that can be remotely connected to global warming is being forced to align with the movement or else.

- *The integrity of science and scientists is decreasing.* Scientists must look out for their own self-interest. They know that if they want to receive research funding, publish papers, and even retain their jobs, they had better support the global warming movement. The net result is a loss of credibility for all scientists and the propagation of falsehoods regarding global warming that are too numerous to count.

- *Personal freedoms are decreasing.* Socialists are using global warming to attack the freedom of speech guaranteed by the First Amendment. Several U.S. states are trying to pass unconstitutional legislation that would make it a crime to speak out against global warming. Socialists are also trying to place all energy sources under government control using a 'smart grid' that will eventually be able to monitor and dictate the energy use of all Americans.

- *Businesses are failing and jobs are being lost.* Hardest hit are businesses related to the energy sector, where hundreds of thousands of jobs have been lost. The coal industry has been driven to the point of total bankruptcy, ruining the economy of states such as West Virginia. The oil and gas industry has been handcuffed, as pipelines and drilling rights are blocked. Higher taxes and regulations are leading to lost jobs for all energy-intensive businesses, which often leave the United States for healthier business climates in developing nations.

- *The climate change agenda is seriously damaging the United States and other free-market economies.* The above examples illustrate that the socialist agenda of crippling affordable energy to weaken the United States is working. Our economy and way of life will continue to be threatened so long as our government bases its policies on the global warming movement.

No one wants to see the Earth's climate destroyed. However, as outlined in the remaining chapters in this book, all of the serious, negative consequences listed above are being inflicted on you to solve a problem that doesn't even exist.

Chapter 2: Greenhouse Gases and the Mechanism for Global Warming

Myth: Greenhouse gases injected into the atmosphere by human activities are causing a catastrophic increase in the Earth's temperature. Specific gases of concern include carbon dioxide (CO_2) emitted from the burning of fossil fuels and methane (CH_4) emitted by cows.

In order to understand whether greenhouse gases are destroying the planet, one must first understand what a greenhouse gas is and the mechanism by which such a gas could potentially impact the temperature of the Earth. At the most fundamental level, the temperature of the Earth's surface and adjacent atmosphere where climate is experienced is controlled by a balance between how much energy enters the surface compared with how much energy leaves the planet and escapes into outer space. This chapter outlines: 1) how visible light from the Sun supplies the energy that controls the Earth's climate, 2) how the heat absorbed from sunlight is used to power all elements of climate, 3) how excess heat is radiated back into outer space to maintain the Earth's temperature, 4) how greenhouse gases impede the escape of radiant heat to promote warming, and 5) the specific impact that carbon dioxide and methane have on this warming effect.

Note: Some concepts and notations used throughout this book may be unfamiliar to those lacking a technical background. Appendices have been included to help readers fill knowledge gaps. Appendix 1 explains scientific notation, orders of magnitude, and Greek prefixes used to describe factors of ten. Appendix 2 provides units and conversion factors for physical quantities such as length, temperature, and concentrations. Appendix 3 is devoted to concepts, units, and conversion factors related to energy. Appendix 4 contains background information regarding light, while Appendix 5 deals with the environmental chemistry of carbon dioxide. Appendix 6 outlines relationships between ice melt and sea levels.

Solar Energy: The Power Source for Earth's Climate

Most of the enormous energy that drives the climate of the Earth originates as light from the Sun.

All phenomena associated with weather and climate consume energy . . . a lot of energy. Energy is required to maintain the temperature of the Earth. Energy is what drives differences in temperature and pressure that control the wind and circulation patterns of air and water around the planet. Energy is required to evaporate all of the water that forms clouds, storm systems, and precipitation. Hurricane Andrew slammed into Florida in 1992 (Fig. 2.1). Even more devastating was the impact of Hurricane Irma during 2017. Irma consumed up to 31 million kilowatt hours of energy (see Appendix 3), which is equivalent to one half of the daily electrical output of the entire United States (see Chapter 8). Where does this massive amount of energy come from?

Fig. 2.1 A composite satellite image compiled by the National Oceanic and Atmospheric Administration (NOAA) to show the progress of Hurricane Andrew across Florida (right) and the Gulf of Mexico[1] in 1992. Images are separated by 24 hour intervals as the storm moved from right to left (east to west).

Weather and climate derive their energy from sunlight. The Sun is constantly bombarding the Earth with light or *electromagnetic radiation* spanning an enormous range of wavelengths and energies[2] (Fig. 2.2) (see Appendix 4). Short wavelength light has a higher energy than light with longer wavelengths. For example, a factor of ten increase in wavelength corresponds to a factor of ten decrease in energy.

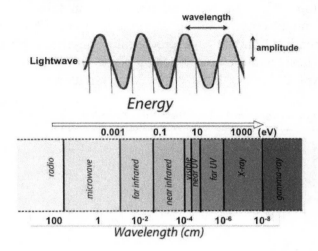

Fig. 2.2 Top – A representation of an electromagnetic wave showing its wavelength and amplitude (strength). Bottom - Various regions of the electromagnetic spectrum showing how different regions are defined on the basis of the wavelength or energy (in electron volts, see Appendix 3) of the light.[2]

Each category of electromagnetic radiation interacts with matter differently. As examples, in moving from high to low energy, x-rays have short wavelengths that allow them to penetrate the body for imaging broken bones. Ultraviolet light has sufficient energy to cause sunburns. Visible light is captured in photovoltaic solar energy devices (see Chapter 8), followed by infrared light used in night vision goggles, microwaves used for cooking, and radio waves used for long-range communication devices.

In terms of climate, the regions of the electromagnetic spectrum of greatest importance involve wavelengths corresponding to incoming visible and outgoing infrared light.

All hot objects emit light having a broad distribution of wavelengths. This light is called *blackbody radiation*. Hotter objects emit light having higher energies and shorter wavelengths. Light emitted by the Sun is dominated by blackbody radiation corresponding to the temperature of its outer atmosphere of 5250°C (9480°F). The distribution of the Sun's blackbody radiation as a function of intensity (or amount of light) and wavelength is depicted in Fig. 2.3.[3] The most intense radiation falls within the wavelength range corresponding to visible light (390 to 750 nm)(see Appendix 2 and 4). The green chlorophyll that plants use for photosynthesis and the color-sensing cones in human eyes both evolved to exploit this dominant visible wavelength regime.

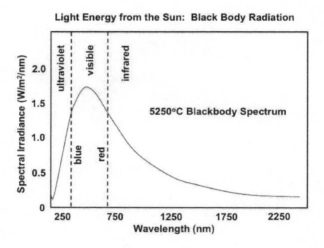

Fig. 2.3 The amount of light energy received by the Earth from the Sun as a function of the wavelength of the incoming light (adapted from Ref. 3). The solid curve corresponds to the light emitted by a black body having the temperature of the Sun. The region of highest light intensity between the dashed lines corresponds to visible light.

The total amount of energy reaching the surface of the Earth from the Sun in the form of light is a staggering 174,000 terawatt[4] (see Appendix 3, Chapter 8), or 6,000 times the rate of energy consumption by Hurricane Irma.

A terawatt is one trillion (10^{12}) watts, or the equivalent energy consumed by ten billion 100-watt light bulbs all burning at once. By comparison, the 47 terawatts reaching us from the Earth's interior[5] (the mantle and molten core) is relatively minor. This chapter outlines what happens to sunlight when it strikes the Earth.

The fact that the energy reaching the Earth from the Sun is not constant, but varies according to internal solar energy cycles and variations in the Earth's orbit, is discussed in Chapter 3.

The Earth's Energy Cycle and The Origins of Weather and Climate

The flow of energy derived from sunlight is extremely complex, as evidenced by the ever-changing weather patterns on Earth.

A simplified diagram showing the net energy flow into and out of the Earth is depicted in Fig. 2.4. In terms of weather and climate, the energy flow involves: 1) the absorption of visible light to create heat, 2) the circulation of that heat throughout all environments, and 3) the conversion of heat into infrared (IR) radiation that is emitted back into outer space.

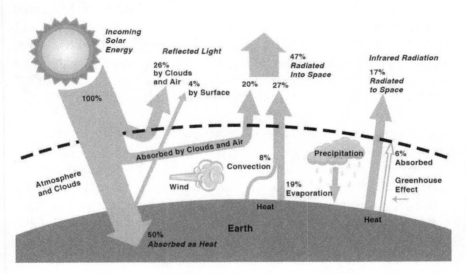

Fig. 2.4 The Earth's Energy Cycle - Incoming solar energy (upper left) follows a complicated trajectory indicated by arrows in the diagram, where arrow sizes scale with the energy involved in a given process. Around 30% of the energy is reflected, clouds and air absorb 20%, while 50% reaches the ground where it is absorbed and converted into heat. This heat powers weather and climate (center). Energy is then radiated back into outer space in the form of infrared radiation, 23% of which originates from the Earth's surface (far right). Global warming is associated with the re-absorption of some of this IR radiation (6%) by greenhouse gases. Adapted from data presented in Ref. 4.

Approximately one third of incoming sunlight (30%) is reflected back into space by clouds, the atmosphere, and the Earth's surface. These reflections transfer no energy to the Earth. Clouds and the atmosphere absorb an additional 20% (one fifth). The remaining 50% (half) of incoming sunlight is absorbed when it reaches the Earth's surface. Upon absorption, the energy contained in sunlight is converted into heat. This heat is what powers the Earth's weather and climate.

The heat absorbed at the Earth's surface is partitioned into three major components. First, heat is responsible for the motions that create air currents, wind, and waves. Imagine how much energy is required to drive all of the wind across the entire planet. Then consider that only 8% of the incoming solar energy (or 16% of the energy absorbed at the surface) is converted into motion. Even more energy is required to evaporate water from the oceans into the atmosphere. Consider how much energy is required to boil a cup of water. Now imagine how much energy is required to produce all clouds on Earth. Although much of this energy is recovered when water vapor condenses and falls as precipitation, the net evaporation-precipitation cycle still consumes around 20% of all solar energy (38% of the light striking the Earth's surface).

Emission of Excess Heat into Outer Space: Infrared Radiation

Almost half of the energy absorbed at the Earth's surface (46%, or 23% of all solar radiation) is not immediately used to generate weather, but is converted back into light that is radiated away from the surface into outer space. However, because the surface of the Earth is much cooler than the Sun (17°C or 63°F compared to the Sun's 5250°C (9480°F)), the wavelengths of the Earth's blackbody radiation are almost twenty times longer than those emitted by the Sun. This means that *most of the energy emitted by the Earth back into outer space is infrared (IR) radiation* that is invisible to the human eye but can be felt at a distance as radiant heat from warm objects. Of this heat, it is estimated that 17% escapes unimpeded. Greenhouse gases temporarily absorb the remaining 6%.

The proposed mechanism for global warming is based on the premise that greenhouse gases injected into the atmosphere by humans are causing more of this radiated heat to be temporarily captured, shifting the Earth's energy balance in favor of heating. Below, this heating mechanism and greenhouse gases are discussed in greater detail.

The Global Warming Mechanism and Greenhouse Gases

Molecules in the atmosphere can absorb infrared (IR) radiation. This absorption makes the atoms in the molecules vibrate faster, which makes them 'hotter.' The molecules 'cool off' by reemitting this IR radiation. Absorption and reemission processes continue until the radiation reaches the upper atmosphere and escapes into outer space. However, repeated absorption and emission slows down the rate at which the IR radiation escapes and makes the atmosphere warmer than it would otherwise be.

Just as materials such as glass are transparent to visible light while others such as black asphalt are opaque, some molecules are highly efficient at capturing IR radiation, while others are not. The two most prominent gases in our atmosphere - nitrogen (78%) and oxygen (21%) - are transparent to IR radiation and do not absorb much of the heat radiated into space by the Earth's surface.

Gases that are not transparent to IR radiation are called greenhouse gases. The impact of a given gas on impeding the escape of the Earth's infrared black body radiation can be quantified by multiplying the ability of a single molecule to absorb the radiation times the relative number of molecules that are present in the atmosphere.

The efficiency with which all major greenhouse gases absorb infrared radiation has been measured using infrared spectroscopy.[6] In infrared spectroscopy, a heat source that emits infrared light is placed at one end of a sample cell of known length containing a specified concentration of a given gas. A detector placed at the other end of the cell measures how much of the emitted light passes through the cell. A recording of the transmitted light, called an infrared spectrum, quantifies exactly how much IR radiation is absorbed as a function of wavelength. A simple expression for this absorbance is described using Beer's Law:[6]

$$A = abc \qquad (2.1)$$

Here **A** is the total absorbance at a given wavelength, **a** is the molecular absorbance (or the absorbance by a single molecule), **b** is the sample thickness, and **c** is the concentration of the gas within the sample volume. The more molecules the light encounters (a product of **b** and **c**), the more IR radiation (or heat) gets absorbed.

The Impact of Carbon Dioxide and Methane on Earth's Climate

The most vilified greenhouse is carbon dioxide, as this gas is tied to fossil fuel emissions. However, the global warming lore often states that methane is 'thirty times worse' in terms of its impact, especially when the goal is to attack the meat industry. Are the claims regarding these greenhouse gases actually justified? Below, such claims are evaluated by examining: 1) how much IR radiation individual greenhouse gases absorb, and 2) how many molecules of each greenhouse gas are present in the Earth's atmosphere.

Infrared absorbance values for the critical greenhouse gases carbon dioxide, methane, and water are shown for all frequencies within the IR spectral window in Fig. 2.5. Water strongly absorbs IR radiation within three broad frequency ranges called absorption bands[7], with the most intense absorption occurring between 3000 and 3700 cm^{-1}. The primary absorption band for carbon dioxide[8] at around 2340 cm^{-1} is more intense, but is also much narrower (i.e. involves far fewer frequencies). The primary absorption for methane at around 3000 cm^{-1} is less intense,[9] but broader than the main absorption band for carbon dioxide. Regions in which no peaks occur are essentially transparent to IR radiation, allowing the IR light to escape into outer space unimpeded.

The total molecular absorbance associated with a given gas is determined by the absorbance at all frequencies, which corresponds to the total area under all peaks in a given spectrum. An analysis of all peaks in the IR spectra in Fig. 2.5 shows that:

On a per molecule basis, the primary greenhouse gas in our atmosphere is not carbon dioxide or methane but water. A single water molecule absorbs over twice the heat of either CO_2 or CH_4, which are actually nearly identical in terms of their ability to absorb infrared radiation.

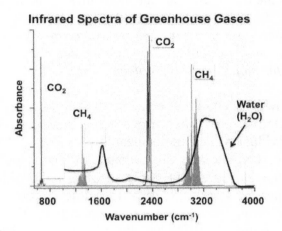

Infrared Spectra of Greenhouse Gases

Fig. 2.5. The absorption of infrared (IR) radiation by greenhouse gas molecules including carbon dioxide (CO_2, sharp open peaks), methane (CH_4, filled peaks), and water (solid dark curve). The wavelength of the IR light is presented in units of cm[-1] (see Appendix 4). Peak heights shown are scaled to approximate the relative molecular absorbance of each gas, and do not represent the actual absorbance values, which differ for each gas experiment. For original data sources, see Refs. 7-9.

The actual effectiveness of water as a greenhouse gas is far greater than the factor of two associated with single molecules. This is because there are many more water molecules present in our near-surface atmosphere than there are of either CO_2 or CH_4.

Over the oceans, which cover 71% of the Earth's surface, the relative humidity is high, approaching 100% or a molecular concentration of 3.2%. Over land the average relative humidity is around 50%, being much higher over tropical jungles and much lower over arid deserts. The average relative humidity over the entire planet[10] represents a molecular concentration of around 2.7% (1/37 of the air or 27,000 parts per million by volume (ppmv))(see Appendix 2 for concentration units). In contrast, the total molecular concentration of CO_2 in the atmosphere is 400 ppmv. Only 400 molecules out of one million (1/2,500, or 0.04%) are carbon dioxide. Methane (CH_4) is present at a concentration of 2 ppmv or only one molecule out of 500,000 air molecules (0.0002%).

On average there are 70 times more water molecules in near-surface air than molecules of carbon dioxide molecules, and 14,000 times more water molecules than molecules of methane.

Based on infrared spectra and concentration information, one can evaluate the relative impact of various greenhouse gases on global warming. Individual water molecules absorb twice the heat of molecules of carbon dioxide or methane. As there are 70 times more water molecules than CO_2 molecules per unit volume, water in the atmosphere absorbs 2 x 70 = 140 times more of the heat radiated by the Earth's surface into outer space than CO_2. Similarly, water absorbs more than 28,000 times the heat absorbed by methane.

On a percentage basis, atmospheric CO_2 absorbs less than 0.74% of the heat absorbed by atmospheric water, while CH_4 absorbs less than 0.004%.

Note that the above numbers include *all* of the CO_2 and CH_4 in the atmosphere rather than the contributions made by humans. If one makes the erroneous assumption that *all* of the 25% increase in atmospheric CO_2 concentrations observed in the Industrial Age is due to the burning of fossil fuels (see Chapter 4), then the maximum human contribution to greenhouse gases that trap heat near the Earth is (0.25)(0.74%) = 0.2% or one part per 500. Such a low value is below our ability to detect it. Measurements performed using three different satellites confirm that increasing CO_2 levels have not caused any decrease in the amount of outgoing IR radiation leaving our planet[11]. In fact, there has been a very slight increase due to a minor increase in sea surface temperatures (see Chapter 6). The contribution made by methane is so ridiculously small that methane will not be discussed in subsequent chapters of this book.

Climate Change advocates do not want to talk about water for obvious reasons. However, when confronted with the fact that water is a greenhouse gas, they reply with two counter arguments.

First, they claim that the above analysis is an oversimplification because atmospheric water concentrations are highly variable. At the Earth's surface, locations can be identified (such as the middle of the Sahara Desert) where the relative humidity is below 0.1%, making CO_2 the dominant greenhouse gas. More importantly, water concentrations in the atmosphere drop dramatically with altitude. At high altitudes, CO_2 and even CH_4 are more concentrated than water. This is because atmospheric temperatures decrease with altitude. Water vapor condenses to form water droplets and eventually ice crystals whenever the temperature drops below 0°C (32°F). For this reason, approximately 50% of all atmospheric water molecules are found within 2 km (1.2 miles) of the surface.[12] No such condensation occurs for carbon dioxide or methane, whose concentrations relative to nitrogen and oxygen stay constant with altitude. However, this observation is irrelevant for two important reasons.

First, the water droplets and ice crystals in clouds ('greenhouse' liquids and solids) are much more efficient than even water vapor at blocking IR radiation. Water is almost 40,000 times more concentrated in liquid water (55 moles/liter) than in humid air (0.0014 moles/liter at 100% relative humidity) (see Appendix 2 for unit descriptions). In fact a water droplet that is only 2 microns thick (1/40 the thickness of a human hair) absorbs more than 99% of the IR radiation within the peaks in its IR spectrum. Clouds inhabit the entire troposphere[12] (the near-surface region where weather occurs) up to altitudes of roughly 12 kilometers (7 miles). The average cloud cover over the entire Earth is 68%. This means that clouds are the dominant factor controlling how much heat escapes our planet in the form of radiant energy, even dwarfing the blockage due to humid air.

Second, although CO_2 and CH_4 concentrations exceed those of water in the upper atmosphere, all gas concentrations in the upper atmosphere are low. Atmospheric pressure drops sharply with altitude. At the edge of the troposphere (around 7 miles), the concentration of all gases is only 10% of what it is at the Earth's surface.[13] Over 99% of the atmospheric mass is contained within 20 miles of the surface. What this means is that essentially all absorption of the heat radiated from Earth occurs close to the surface where both water vapor and clouds are prevalent. The outer atmosphere has a negligible impact on the Earth's temperature.

Climate change advocates rely on a second major argument to try to rebut the fact that the increase in CO_2 concentrations due to human activities is negligible relative to the impact of water. This argument states that because the amount of outgoing IR radiation (6% of the solar energy total) is so huge, even the smallest increase in the net greenhouse gas concentration can have a large impact on the Earth's climate. The validity of this claim is examined in the next two chapters by: 1) comparing the modern net change in greenhouse gas concentrations with temperature changes known to be associated with natural causes (Chapter 3), and 2) making direct head-to-head comparisons between CO_2 concentrations and the Earth's temperature over time scales ranging from a billion years ago to the present (Chapter 4).

Summary: Water, both in water vapor and in clouds, is by far the most important greenhouse substance on Earth, dwarfing any effects attributed to greenhouse gases such as carbon dioxide and methane. The concentrations of carbon dioxide and methane in our current atmosphere are so low that they have a negligible impact on our climate, let alone the concentrations of these gases that are attributable to human activities. Humans can continue to burn fossil fuels and eat meat without worrying about destroying the planet due to global warming.

Chapter 3: Natural Factors that Control the Earth's Climate

Myth: The temperature of the Earth was essentially constant until humans started burning fossil fuels to trigger runaway Global Warming.

When Al Gore's movie *An Inconvenient Truth* was released in the 1990s, the claim was made that the Earth's climate had been warming since 1900 by over 1°C. Progressives attribute this warming to alarming increases in atmospheric carbon dioxide released in the utilization of fossil fuels by the world's energy economies. However, as water is by far the dominant greenhouse gas (Chapter 2), it is not immediately apparent how the miniscule contribution that humans make to the total greenhouse gas concentration can be controlling the climate. How much has the Earth's temperature actually changed in modern times (Chapter 6)? Are these changes normal or unusual? To answer these questions, one must first have an appreciation for changes in climate known to be produced by purely natural causes.

It is important to recognize that the Earth's climate has been continuously changing during its entire 4.6 billion year history. Earth's climate has never been constant.

Climate change actually encompasses an enormous range of temperatures and time scales. Below, these changes are highlighted based on natural mechanisms that occur over time scales ranging from millions to tens of years.

Continental Drift: Climate Change Over Millions of Years

Continental positions determine the distribution and circulation of heat on Earth and have a major impact on our planet's long-term climate. Current continental positions are promoting a relatively cold era of recurring Ice Ages.

The largest climate changes transpire over time periods of 20 to 100 million years[1] (Fig. 3.1). These changes (both gradual and catastrophic) are associated with continental motions due to plate tectonics or continental drift.[2]

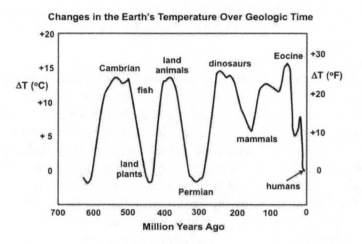

Fig. 3.1 Estimated average global temperatures during the past 700 million years based on the geologic record (from data presented in Ref. 1). Geologic eras are indicated in capitalized labels. Lower case labels refer to life forms that appeared during a given era.

These changes have absolutely nothing to do with carbon dioxide levels in the atmosphere (see Chapter 4). The coldest known climatic episodes or ice ages occurred within the aptly named Cryogenian Period[3] (850 – 580 million years ago (Mya)). During the worst known ice age (the Mironoan Period (850 – 650 Mya), it is thought that most of the Earth's oceans were frozen to a depth of up to a mile, while continental temperatures near the Equator resembled those in modern Antarctica (averaging -50°F or -45°C). At the other extreme, there have been several extended periods when average temperatures exceeded 90°F (32°C). One such warm era was the Cambrian Period[4] (550 – 505 Mya), which saw the creation of more new animal species than any other time in Earth's history. One of the coolest periods prior to today was the Permian Period, which coincided with the largest mass extinction in Earth's history. In fact, most of Earth's mass extinctions occur during

30

periods of global cooling rather than global warming (Chapter 7). The most recent era of warm climates occurred in the Eocene Epoch,[5] during which tropical rain forests flourished above the Arctic Circle. Since the Eocene ended 34 Mya, the Earth has experienced cooler climates.

The positions of the continents have a major impact on long-range climate because they determine heat circulation patterns in both our oceans and our atmosphere. Sometimes the continents are near the equator. Sometimes continents are near the poles. Sometimes the continents merge into a single landmass, while at other times widely spaced continents are present. For much of the time since animals first appeared on land 400 million years ago (less than 10% of Earth's history), there were no major ice caps on Earth. The current epoch of periodic Ice Ages started only 2.5 million years ago due to the formation of a land bridge between North and South America.[6] This continental fusion created a new oceanic circulation pattern that sends warm water from the tropics up into the continent-rich regions of the Northern Hemisphere. The air above these warmer waters is laden with atmospheric moisture. If sufficient moisture condenses to form snow over cold continental masses, and if this snow accumulates faster that it melts, glaciers form. Glaciers have a high albedo (or ability to reflect sunlight back into outer space, see Chapter 2), ultimately resulting in a cooler climate. Unless more sunlight starts to fall on the colder polar region, the cooling of the continents promotes the spreading of glacial masses, resulting in an Ice Age.

Variations in the Earth's Orbit: Recurring Ice Ages (2.5 Million to 10,000 Years Ago)

Periodic changes in the Earth's orbit influence how the energy that the Earth receives from the Sun is distributed, resulting in our current era of recurring Ice Ages. The Earth is currently experiencing the high temperature end of the latest Ice Age cycle.

The periodic ice ages that define the Earth's current climate highlight the second most important natural climate change mechanism, which involves how much sunlight the poles receive relative to the equator during different seasons of the year.

Glaciation occurs during extended periods when the temperature difference between the equator and the poles is the greatest, pumping the maximum equatorial moisture to colder regions where it falls as snow and is retained. Ice ages end when cooler equatorial waters pump less moisture to heated regions at the poles, resulting in lower snow levels and warmer landmasses that melt the glacial masses. This mechanism operates on time scales of tens to hundreds of thousands of years compared with the millions of years required for continental drift. During the current Ice Age era, this mechanism (rather than swings in atmospheric carbon dioxide levels, see Chapter 4) has accounted for shifts in the Earth's temperature of up to 20°F (10°C) compared with the larger 30°F (15°C) swings that have accompanied continental drift.

Ice ages do not occur in a random fashion, but follow a specific and repeatable pattern (Fig. 3.2).

Earthly Temperatures Inferred from the Vostok Ice Cores

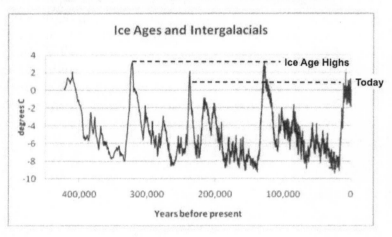

Fig. 3.2 Estimated relative global temperatures over the past 500,000 years based on analyses of the Vostok ice cores (adapted from data in Ref. 7). The temperatures through Ice Age exhibit a regular saw-tooth pattern that repeats every 100,000 years due to orbital Milankovitch cycles.

Both deep-sea sediment and ice core samples show that ice ages take place every 22,000 years.[8] Each ice age consists of a gradual cooling period and the growth of massive polar ice sheets followed by a period of rapid melting. The episodes having the greatest temperature differences between the cooling and heating periods (typically 15°F or 8°C) occur every 100,000 years. *We are currently within the high-temperature end of the 24ᵗʰ of these modern major ice age cycles.* Based on known patterns, the Earth is predicted to descend into the next major ice age in less than 5,000 years. *This major cooling event that is on the horizon is the climate change event that humans should really be worried about.*

Every year, the Earth experiences a cycle of 'global warming' followed by 'global cooling' that accompanies the seasons. In much of the Northern Hemisphere, the 'climate change' experienced between the summer and winter months amounts to around 50°F (not 1°F!). The same orbital parameters that give rise to the seasons exhibit long-term periodic variations. These variations are called *Milankovitch cycles.*

Milankovitch Cycles

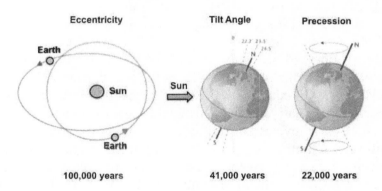

Fig. 3.3 A depiction of the three Milankovitch cycles associated with variations in the Earth's orbit around the Sun. Left – The annual orbit of the Earth around the Sun cycles between a circular and an elliptical shape every 100,000 years. Center – The angle of the Earth's axis relative to the Sun cycles between 22.2° and 24.5° every 41,000 years. Right – The Earth's axis wobbles around a tilt angle of 0° in a cycle that requires 22,000 years to complete. At one end of the cycle, the North Pole faces the Sun in the winter, while at the other end, the North Pole faces the Sun in the summer.

The pattern of ice ages, both in terms of duration and severity, are highly predictable based on Milankovitch cycles.[9] These cycles involve periodic changes in the position of the Earth relative to the Sun, controlling how much sunlight each region of the globe receives. Milankovitch cycles consist of three important interactive components (Fig. 3.3).

First, the Earth's orbit around the Sun is neither perfect nor unchanging, but slowly oscillates between being elliptical to nearly circular on a time scale of around 100,000 years. When the orbit is elliptical, the Earth is 2.5% closer to the Sun and receives more heat during two seasons of the year. It is 2.5% farther away and receives less heat during the other two seasons.

Second, the Earth's axis is tilted relative to its orbit (currently at an angle of 23.5°). This tilt angle increases and decreases with a periodicity of 41,000 years. Smaller tilt angles decrease the differences between the seasons, but increase the temperature difference between the Equator and the Poles.

Third, the tilt of the Earth's axis is not always angled in the same direction. Like a spinning top that is slowing down, the axis wobbles or undergoes precession with a periodicity of 22,000 years. At one end of the cycle, the North Pole faces the Sun in the winter, while at the other end it faces the Sun during the summer.

All three Milankovitch cycles influence how much sunlight the poles receive relative to the equator during each season of the year. When all three cycles are in phase once every 100,000 years, they reinforce each other to either stimulate massive glaciation or rapid ice cap melting. When the cycles are out of phase, they partially cancel each other out to smooth out climatic variations. The net effect is a complex saw-tooth pattern that generates highly predictable and periodic ice ages.

Climate Throughout Human History

The energy produced by the Sun is not constant. Solar output exhibits both long- and short-term variability. The Earth is currently experiencing a warm era associated with long-term solar cycles and a transition period in short-term cycles.

Most of the warming and cooling trends observed during human history are related to a third periodic factor influencing our climate. This factor operates on times scales of ten to a thousand years and results in temperature shifts spanning a total range of around 7°F (4°C). These shifts arise from the fact that *the output of energy and radiation from our Sun is not constant, but changes according to both long-term and short-term cycles of solar activity*. These solar cycles, and their connection with the Earth's climate, have been documented using the recorded history of sunspot cycles, aurora observations, radiocarbon dating techniques, and changes in solar radiance. Below, the climate changes that have occurred during human history are highlighted, followed by a brief discussion of how scientists have correlated these changes with variations in solar activity over the past 5,000 years.

There is a myriad of evidence based on soil sample analyses, analyses of dead vegetation, sediment analyses, glacial ice cores and moraines, and human artifacts to show that significant climatic variations have been observed throughout human history (Fig. 3.4). (See Chapter 7 for the impact of these changes on humanity.)

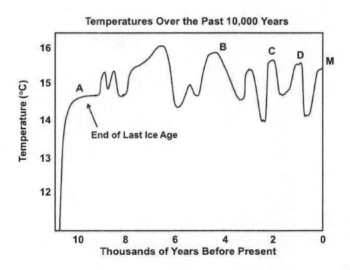

Fig. 3.4 Temperature records going back 11,000 years based on ice core data from Greenland (adapted from data in Ref. 10). The temperature axis is in °C. The letters A-D correspond to the maps of Icelandic glaciers depicted in Fig. 3.5, while the letter M stands for modern times.

Historical Glaciation of Iceland

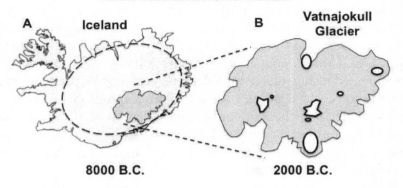

A   Iceland

8000 B.C.

B   Vatnajokull
    Glacier

2000 B.C.

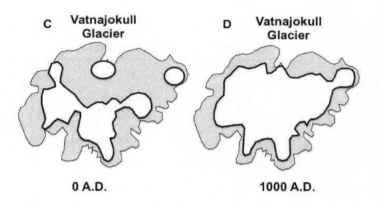

C   Vatnajokull
    Glacier

0 A.D.

D   Vatnajokull
    Glacier

1000 A.D.

Fig. 3.5 Some of the estimated positions of the Vatnajokull Glacier in Iceland during the past 10,000 years. (Adapted from Ref. 11.) The modern extent of the glacier (outlined in gray) is included in all maps as a point of reference. In map A (8,000 B.C.) the entire island of Iceland is shown, with the dashed line indicating that glaciation encompassed almost the entire island. In maps B-D, the background shown is that of modern day Vatnajokull glacier, with white regions indicating glaciated areas. Note that major advances and retreats occurred between B and C as well as C and D (see Fig. 3.4).

Over 10,000 years ago, when cave men roamed the planet near the end of the last Ice Age, Iceland was completely submerged by a polar ice cap that was so thick that its weight pushed the island below sea level. At the peak of the Egyptian civilization (3,000-5,000 years ago when the Great Pyramids were built), there was almost no ice left in Iceland at all. Obviously, human activities did not make all of this ice disappear. Before the birth of Christ 2,300 years ago, Iceland was once again covered by glacial ice. However, by the time the Vikings first settled Iceland around 930 A.D. in the early stages of the Medieval Warm Period[12] (Fig. 3.6), much of this ice was gone.

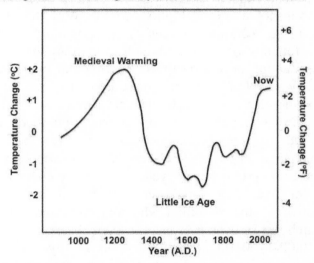

Fig. 3.6 A high resolution graph showing global temperatures over the past 1,200 years adapted from the IPCC compilation shown in Fig. 1.3 (Ref. 21, Chapter 1).

Vatnajokull Glacier was smaller then than it is today. The glacier continued to melt during much of the remaining Medieval Warm Period, during which time the Vikings established farming communities in Greenland. Unfortunately for the Vikings, during the subsequent Little Ice Age between 1350 and 1900, all Icelandic glaciers advanced until they covered 1/3 of Iceland. Due to severe cold, the Vikings started to leave Greenland[13] in 1350. Greenland was abandoned by 1500. The Vikings almost evacuated Iceland in the 1700s. Since 1900, the glaciers have once again been retreating due to increasing climactic temperatures. Today, Vatnajokull covers 12% of the island, which is more ice than was present at the time of the Vikings. However, today's climate is cooler than it was during much of the Medieval Warm Period (Fig. 3.6).

Solar Irradiance: The Mechanism for Historical Climate Change

Most of the climatic changes observed during human history can be traced back to changes in solar activity. The fact that the Sun's activity is variable was known even to ancient civilizations. The first clues regarding variable solar activity came from observations of sunspots. The Chinese have records of sunspot activity based on naked eye observations as early as 58 B.C.[14] Even then, people knew that sunspot activity is highly variable. With the advent of the telescope, more accurate readings of sunspot numbers have been taken as a function of time. Although maximum sunspot numbers vary, sunspots appear and disappear in a regular cycle that repeats every 11.2 years. The Sun's magnetic poles flip after each cycle, resulting in a solar magnetic field cycle of 22 years.[15] Long-term variations in average maximum sunspot numbers are more complex, but appear to be the sum of other periodic solar phenomena including the Gleisberg (88 years), DeVries (208 years), and Eddy (1000 years) cycles.[15] Intense sunspot activity can continue for over a hundred years, followed by equally long periods, such as the so-called Maunder minimum (1645-1715), during which few or even no sunspots were observed.

From 2800-1700 B.C., when sunspot activity was high, there was little ice in Iceland. From 1500-200 B.C., when sunspot activity was low, Iceland was heavily glaciated. The strong correlation between sunspot activity and glaciation is not a coincidence, but reflects the fact that high sunspot activity is an indicator of increased solar activity. Connections between sunspots and solar activity are apparent in changes in the stream of electrons, protons, and alpha particles emitted by the Sun called the solar wind.[16]. Increases in the intensity of the solar wind have always been observable in the form of auroras and more recently in the disruption of radio communications and electromagnetic devices. Both ancient and modern observations show that there is a direct correlation between the number and intensity of auroras and the sunspot cycle. Solar activity as indicated by the solar wind is the highest when sunspot activity is also highest.

A more quantitative description of variations in the intensity of the solar wind has been established based on observations of cosmic rays.[17] Cosmic rays from interstellar space react with nitrogen atoms in the atmosphere to produce the radioactive ^{14}C isotope of carbon. This carbon is incorporated into carbon dioxide (CO_2) that plants absorb via respiration. Radiocarbon dating of tree rings and other plant matter is based on measuring the relative amounts of ^{14}C (which decays with time) and ^{12}C (which does not) in a given sample of organic carbon.

Radiocarbon analyses reveal that baseline ^{14}C levels are variable, which means that the irradiation of the Earth by cosmic rays has also been variable. It is now known that this variability is directly related to the strength of the solar wind, which sets up a magnetic field that partially blocks the cosmic rays. Variations in ^{14}C levels are a direct measure of variations in solar activity as indicated by the strength of the solar wind. The lower ^{14}C levels are, the stronger the Sun's activity was at the time the plant matter containing the ^{14}C was created. This agreement has allowed scientists to use ^{14}C in tree rings to track solar activity from the present back to 5500 B.C. (7500 years ago).

Since the beginning of human history, there has been a good match between our climate and solar activity. Solar activity from 900 A.D. to the present is shown in Fig. 3.7.[18] Although not all peaks and valleys in solar activity are reflected in estimates of the Earth's average climate, the major features are in agreement (see Fig. 3.6). Solar activity was high from before 900 to 1300, including the Medieval Maximum. This period coincides with the Medieval Warm Period when the Vikings occupied Greenland and Iceland. Solar activity was generally low between 1300 and 1700, including the Wolf, Sporer, and Maunder Minima. This low solar activity is coincident with the Little Ice Age. Solar activity has oscillated and generally increased since then to reach the Modern Maximum around 2000.

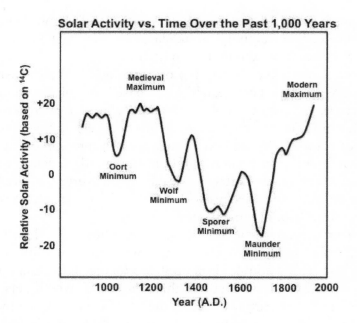

Solar Activity vs. Time Over the Past 1,000 Years

Fig. 3.7 Relative activity of the Sun based on analyses of the ^{14}C content of plant matter (adapted from data in Ref. 18). In recent history, solar activity was the highest during the Medieval Maximum and lowest during the Maunder Minimum. Notice the close comparison between the solar activity shown here and the estimated global temperatures shown in Fig. 3.6.

Changes in average global temperatures since 1900 (Fig. 3.8[19]) are much more consistent with oscillations in solar activity and the average amount of energy that we receive from the Sun than they are with the exponential increase in fossil fuel emissions (see Chapters 4, 6, and 9). Note that the Earth's temperature increased from 1880 to 1935, decreased from 1935 to 1980, increased from 1980 to 1990, and has since leveled off. The temperature did not continuously and dramatically increase to mirror the increasing CO_2 emissions depicted in Fig. 1.1. The most recent measurements show that sunspot numbers may be starting to decrease, which could eventually result in a cooling trend.

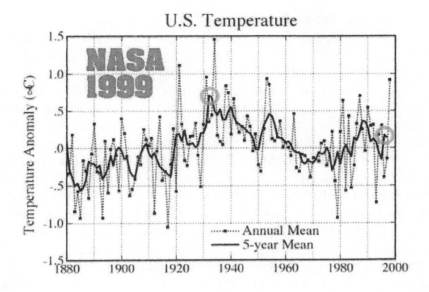

Fig. 3.8 Average temperatures (in °C) in the United States between 1880 and 2000 tabulated by the National Aeronautic and Space Administration in 1999.[19] This temperature compilation clearly shows the warming and cooling cycles that have occurred in recent history, including the rise in temperatures between 1890 and 1940, the drop in temperatures between 1940 and 1975, and the increase in temperatures from 1980 to the present. Note that current temperatures are cooler now than they were in the 1930s (compare positions of small circles). (Note: This 1999 NASA compilation was produced prior to the era of extensive government tampering with archival climate data (see Chapter 9).)

Changes in Solar Activity vs. Greenhouse Gas Concentrations

The link between solar activity and climate is now being measured directly by detecting changes in the total amount of sunlight impinging on the Earth. Satellites have been in place since 1976 that continually measure the average solar irradiance over the course of an entire year.[20] The average value of 1.361 kilowatts per square meter (kW/m^2, see Chapter 8) fluctuates by 6.9% during the year from 1.412 (in January) to 1.321 (in July) kW/m^2 due to changes in the distance between the Earth and the Sun. At exactly the same time of year, the solar irradiance typically varies by 0.1%, although differences as high as 0.3% have been seen over the course of sunspot cycles.

Global warming advocates argue that changes in solar irradiance by 0.3% are too small to account for any warming seen in modern times, and are negligible relative to the 6% of the outgoing energy from Earth that is blocked by greenhouse gases. However, if one considers the blockage due to man-made CO_2 emissions, the exact opposite is true. The solar irradiance represents *all* of the energy coming into the Earth. *All* greenhouse gases are responsible for impeding 6% of that energy that the surface radiates back toward space. However, as water is by far the dominant greenhouse gas (see Chapter 2), the entire 400 ppmv of CO_2 in the atmosphere accounts for only 0.25% of the greenhouse effect. This means that the total contribution that CO_2 makes to the Earth's energy balance is $(6\%)(0.0025) = 0.015\%$. If one assumes that *all* of the CO_2 increase of 20 ppmv since satellite measurements of solar irradiance have been taken is due to the burning of fossil fuels, a maximum of 5% of the CO_2 in the atmosphere in that time span can be attributed to humans. This makes the human contribution to our net energy balance $(0.0125\%)(0.05) = 0.00075\%$. *If variations in incoming solar irradiance of 0.1-0.3% are 'negligible' as claimed, what does that say about the contribution that human induced increases in CO_2 levels make to the Earth's energy balance, which are 400 times smaller than that?*

In closing this section on the Sun, climate change advocates will be happy to know that their prediction of a fiery end to the Earth is correct. This is because the energy output from the Sun is slowly increasing. The primary source of light and energy within the Sun is the fusion reaction that forms helium atoms from hydrogen atoms (see Chapter 8). Over millions of years, this fusion reaction is changing the size, composition, and energy output of the Sun.[21] At the current stage in the Sun's life cycle, the luminosity of the Sun has been increasing by 1% every hundred million years. This means that since the end of the Cambrian explosion of life on Earth 500 million years ago, the amount of sunlight striking the Earth has increased by 5%. As the life cycle of the Sun continues, the Sun will eventually become hot enough to evaporate the oceans of Earth and incinerate all life on the planet.

Random events including volcanic eruptions and asteroid impacts represent a final factor that can have short-lived yet catastrophic climactic consequences for life on Earth (see Chapter 7). The largest volcanic eruption during the past 10,000 years[22] occurred in Tambora, Indonesia in 1815. This eruption was 200 times larger than the Mount St. Helens eruption. It spewed forth up to 40 cubic miles (170 cubic kilometers or km^3) of material, injecting volcanic ash and an aerosol of sulfuric acid droplets into the upper atmosphere that influenced the global climate for over a year. The volcano rapidly induced a planetary temperature drop of up to 5°F (3°C), producing near freezing temperatures in the tropics, and crop failures that caused over 100,000 people to starve to death. Tambora was puny compared with the eruptions of mega-volcanoes such as the one in Yellowstone National Park that ejected 600 cubic miles (2500 km^3) of volcanic materials onto North America 2.1 million years ago. The Toba eruption in Sumatra that occurred 74,000 years ago ejected 670 cubic miles (2800 km^3) of volcanic materials, making it 2,800 times larger than the Mount St. Helens eruption. It is thought that the Toba eruption cooled the Earth's temperature by up to 9°F (5°C) for several years. Based on genetic evidence, some scientists believe that this eruption reduced the Earth's total human population down to between 3,000 and 10,000 individuals, pushing early humanity to the brink of extinction.

The largest volcanic events in Earth's history have occurred over longer periods of time due to the motions of the continents or plate tectonics (see above). The largest of these eruptions took place over a period of 65,000 to one million years around 250 million years ago to form the Siberian Traps.[23] This episode of eruptions covered 770,000 square miles (the size of Europe) with a total amount of 500,000 cubic miles of lava. Scientists believe that this eruptive period is responsible for the largest mass extinction in Earth's history during the Permian Period (see Chapter 7), during which 95% of all plant and animal species disappeared. A similar eruption occurred at the end of the Cretaceous Period 66 million years ago to form the Deccan Traps that cover half of India.

Some scientists believe that the Deccan Traps eruption contributed to the extinction of the dinosaurs.[24] However, another popular theory is that the mass extinctions observed at the boundary between the Cretaceous and Tertiary Periods (the K-T boundary) was caused by a 6-mile diameter asteroid or comet that slammed into the Yucatan Peninsula in Mexico 65 million years ago.[25] This asteroid hit the Earth at a speed of one hundred thousand miles an hour. The resulting impact was equivalent to the explosion of one hundred million megatons of TNT, creating the Chicxulub impact crater that is 110 miles in diameter. Materials injected into the atmosphere from the impact blocked up to 90% of the sunlight reaching the Earth for as long as several months, plunging the Earth into a global winter. The net result was the extinction of the dinosaurs and as many as half of the plants and animals on Earth.

Global warming advocates would like you to believe that the use of fossil fuels is an environmental disaster equivalent to the Chicxulub impact. Fortunately for humans, the only real 'disaster' is the extent to which the media, our schools, and climate change scientists have ignored, modified, and falsified climate data to promote a purely political agenda. The Vikings and other ancient human populations would be surprised to learn that they never lived through the Medieval Warm Period or the Little Ice Age. The cooling period that led *Time* magazine in 1977 to warn of an impending Ice Age (see Chapter 1) absolutely positively never happened. Instead, everyone is supposed to believe that the Earth's temperature was always constant, but is now increasing by up to several degrees per year due to fossil fuel emissions. It just isn't true (Chapter 6). This extensive rewriting of climate history represents the most pervasive and damaging example of scientific fraud in the history of mankind (Chapter 9).

Summary: The climate of the Earth has been constantly changing since the dawn of time. Variations in our planet's average temperature due to natural causes have ranged from over 90°F down to freezing temperatures, or a span of over 60°F. Most of the periodic temperature increases and decreases observed in human history are consistent with variations in the output of energy from our Sun. The mild heating and cooling periods seen since 1900 (each less than 2°F) reflect changes in solar activity rather than runaway global warming.

Chapter 4: The Impact of Atmospheric Carbon Dioxide on the Earth's Climate

Myth: Carbon dioxide (CO_2) levels in our atmosphere are as high as they have ever been due to the burning of fossil fuels. As CO_2 concentrations control the temperature of the Earth, fossil fuel burning is leading to catastrophic global warming.

No one disputes that at current rates, the burning of fossil fuels introduces 33 billion tons of carbon dioxide into the atmosphere every year.[1] Climate Change advocates argue that the emission of billions of tons of this greenhouse gas must be destroying the planet due to global warming. They would have you believe that humans control atmospheric CO_2 levels, and that fossil fuel emissions have driven CO_2 concentrations higher than they have ever been. Under President Obama, the Environmental Protection Agency (EPA) went so far as to classify CO_2 as a toxic pollutant[2] that is destroying all life on Earth (see Chapter 7) as a rationale to try to close all coal fired power plants in the United States. This chapter explores whether any of the above claims are valid by highlighting what is known about atmospheric carbon dioxide based on the geologic record and the Earth's carbon cycle.

Carbon Dioxide Levels over Geologic Time

Climate Change advocates do not want anyone to bring up the geological history of the Earth. They only want you to focus on the last one hundred of the 4.6 billion years of Earth's existence, and to ignore everything that happened prior to the current Industrial Age. The reason for this is that the geologic record does not support their primary claim that rising CO_2 concentrations control the Earth's climate. In this section, we highlight the geologic record to evaluate some of the arguments that provide the underpinnings of the Global Warming movement. To provide perspective for this chapter, the current CO_2 concentration in the atmosphere is 400 parts per million[3] (ppmv) (see Appendix 2 for explanations of concentration

45

units). Carbon dioxide accounts for 0.04% of the molecules (one molecule out of every 2,500) and 0.06% of the mass (or 613 ppmw) present in our current atmosphere. The total weight of CO_2 in the atmosphere is 3,400 Gton (G = giga = 10^9 or one billion tons) compared with the total atmospheric mass of 5.7 million Gton.

Are carbon dioxide levels as high as they have ever been? The geologic record, involving an analysis of the abundances of carbonate minerals, coal and oil deposits, and fossils throughout the history of the Earth, reveals that the exact opposite is true. Carbon dioxide has always been in the Earth's atmosphere. In fact, CO_2 was the second most prevalent atmospheric gas over 3 billion years ago, surpassed only by nitrogen. Oxygen concentrations were negligible.

Primeval concentrations of carbon dioxide[4] were as high as 20%, or almost 300,000 ppm. These CO_2 concentrations are 500 times greater than modern levels.

Carbon dioxide concentrations remained high until bacteria evolved whose metabolism was based on aerobic photosynthesis. Photosynthesis, on which most modern plant life is based, consumes carbon dioxide, water, and sunlight to create organic matter (see Chapter 7). Oxygen gas (O_2) is released as a waste product. By 2.5 billion years ago, the conversion of CO_2 into O_2 had progressed to the point where oxygen started to accumulate in the atmosphere. The build-up of O_2 actually led to Earth's first mass extinction, as the microorganisms producing O_2 finally started to choke on their own emissions.[5] Eukaryotic cells (the basis for animal life) evolved a new metabolic system to adapt to and exploit the changing atmospheric conditions,[5] reacting oxygen with organic matter (food) to produce energy and generate CO_2 as a waste product (see the Biological Carbon Cycle below).

From 2 billion to 700 million years ago, carbon dioxide levels continued to drop. Estimates of atmospheric concentrations of CO_2 over the past 700 million years are summarized in Fig. 4.1,[6] along with temperature data covering the same period. By the Cambrian Period (550 million years ago) when the largest explosion of new animal species occurred, *CO_2 concentrations were still as high as 6500-7000 ppmv, or 17 times higher than they are today.* Although there were periodic fluctuations in CO_2 levels, CO_2 concentrations gradually decreased from the Cambrian to the Permian Period (300 to 250 million years ago) to only 250 ppmv. For the next 100 million years or so, concentrations increased to a high of around 2200 ppmv (almost 6 times modern values) during the Cretaceous Period when dinosaurs roamed the Earth.

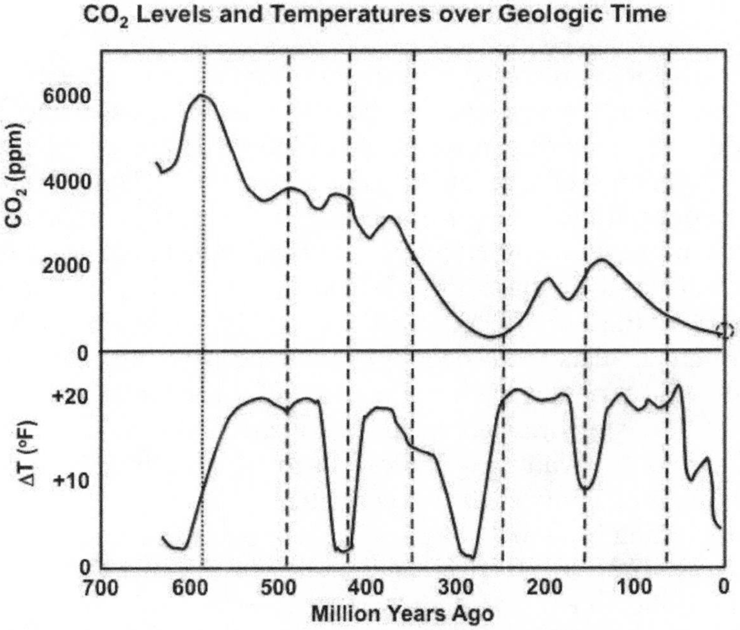

CO_2 Levels and Temperatures over Geologic Time

Fig. 4.1 A direct comparison between atmospheric carbon dioxide concentrations (in ppmv) and average global temperatures (in °F) over the past 600 million years of geologic history. (Adapted from Ref. 6.) Notice that there is absolutely no resemblance between the two curves.

It should be clear by now that the Earth is not experiencing all-time high CO_2 levels as claimed. The opposite is true.

Since the Cretaceous Period, CO_2 levels have continued to drop to all-time lows throughout the current era of recurring Ice Ages.

During the most recent Ice Ages (see Oceanic Sources and Sinks below), concentrations have fluctuated between 180 ppmv and 330 ppmv.[7] In fact, *if CO_2 levels were to drop much lower, all plant life and the entire food chain on Earth would be endangered* (see

47

Chapter 7). Since the end of the last Ice Age 10,000 years ago, CO_2 levels have increased from 180 ppmv up to the current value of 400 ppmv. Almost half of this increase (from 180 ppmv to 280 ppmv) occurred prior to 1880. Note that the increase in CO_2 levels during human history is a minor blip (see the circle in Fig. 4.1) compared with the changes that have occurred over geologic time.

A Direct Comparison between CO_2 Levels and Climate

The geologic record is critical because it reveals *actual* relationships between carbon dioxide levels, climate, and life on Earth. Comparisons can then be made between those relationships and the claims of Climate Change advocates. Claims include: 1) CO_2 concentrations control the Earth's temperature, and 2) the recent increase in temperature of the Earth by as much as 2°F is due to a 120 ppm increase in CO_2 concentrations that is exclusively caused by the burning of fossil fuels. With these claims in mind, environmentalists want people to imagine how dire conditions on Earth will become if the burning of fossil fuels continues. However, the geologic record can be used to check the validity of these predictions by evaluating the extent to which climates of the past correlate with known CO_2 concentrations.

According to global warming advocates, temperatures should go up when CO_2 levels rise, and go down when CO_2 levels fall, tracking the CO_2 curve shown in the top of Fig. 4.1. Using their assumption that the Earth's temperature should increase by 2°F (1°C) for every 100 ppmv-increase in CO_2 concentration, the temperature during the age of the dinosaurs (250 million years ago), when CO_2 concentrations were 2,200 ppmv should have been 36°F or 20°C higher than today. In the Cambrian Period (550 million years ago), when CO_2 levels were 6,500 ppmv, the temperature should have been over 180°F (82°C), or almost hot enough to boil water. Three billion years ago and earlier, before primeval CO_2 levels started to drop, Earth should have been so hot that all water would have boiled away, and oceans would not even exist. The Earth would have been a fiery hell like the planet Venus.

In fact, Climate Change advocates are constantly warning that we are approaching the 'tipping point' beyond which the Earth's climate will descend to a state resembling that of Venus. This claim is absolutely ridiculous. For one thing, Venus is one third closer to the Sun than the Earth, receives twice the solar irradiance as the Earth, and would be hot enough to boil away the oceans on the basis of its orbit alone. More importantly, the high-pressure atmosphere of Venus is composed almost entirely of carbon dioxide (96.5% CO_2, 3.5% nitrogen, and trace gases such as water (at 20 ppmv)).

The CO_2 concentration in the Venusian atmosphere is 30,000 times greater than that found on Earth, equivalent to 12 million ppmv in our atmosphere. Given the differences in CO_2 concentrations and orbital positions, the greenhouse effect associated with CO_2 is 60,000 times greater on Venus than it is on Earth. No possible scenario exists by which the Earth can come remotely close to replicating the greenhouse effect experienced by Venus (see Fossil Fuels below).

How do the above predictions compare with what we know about global temperatures based on the geologic record (see Chapter 3)? Our oceans are known to be at least 4 billion years old. Life inhabited the oceans at least 3 billion years ago when Climate Change predictions claim that Earth should still have been too hot for oceans to even exist. Over 2.5 billion years ago, while oceanic photosynthetic organisms were busily lowering atmospheric CO_2 concentrations by converting CO_2 into O_2, temperatures should have still exceeded the boiling point of water. Based on CO_2 concentrations alone (exceeding 7,000 ppmv), the planet should have been a hothouse during the Cryogenian Period 800 million years ago (Mya) when the Earth was actually suffering through the most massive Ice Age in its history (see Chapter 3). Hmm.

Between 700 million years ago and the present (the era of multicellular animals), direct and continuous comparisons can be made between the average temperature of the Earth (see Chapter 3) and CO_2 levels. In Fig. 4.1,[6] CO_2 concentrations are represented by the top curve and corresponding temperatures appear below. Most of the vertical lines in Fig. 4.1, included to help guide the eye, represent some of the major mass extinctions in Earth's history.

Imagine that you are in a time machine. By starting in the present and going back in time, you should experience increasing temperatures corresponding to increasing CO_2 concentrations for much of this 700 million year time period. Going back in time for the first 40 million years, temperatures increase by 7°C (13°F) per each 100 ppmv increase in CO_2 concentrations, exceeding even the direst global warming predictions by more than a factor of three. However, the maximum temperature reached by the Eocene (40 Mya) is never again exceeded during the next 650 million years. In fact, continuing back in time for the next 100 million years (from 50 to 150 million years ago (Mya)), *temperatures flatten and then drop by 10°F (6°C) even though CO_2 levels continue to increase by a factor of two.* Conversely, moving back from 150 – 175 Mya, *while CO_2 levels are dropping, temperatures increase by 10°F.* From then back to 700 Mya, temperatures continue to cycle between warming and cooling periods every 100-150 million years or so even though atmospheric CO_2 concentrations increase by over 6000 ppmv (50 times the increase seen in modern times) during the same time span. The bottom line is that:

Over billions of years, the geologic record clearly shows that there is no long-term correlation between atmospheric CO_2 levels and the Earth's climate. There have been periods in Earth's history when CO_2 concentrations were over 15 times higher than they are today, yet temperatures were identical to or even colder than modern times. The premise that CO_2 emissions are causing catastrophic global warming is a total myth

Factors Controlling Atmospheric CO_2 Concentrations

The claim that fossil fuel emissions control atmospheric CO_2 concentrations is also invalid. What made carbon dioxide appear and disappear during the geologic history of our planet? Clearly, humans had nothing to do with any of the data presented in Fig. 4.1. These large fluctuations in CO_2 levels must have been due to natural causes. What are these causes, and how do they compare with changes induced by humans? Below, several key issues involving carbon dioxide are highlighted in the context of each of the major sources and sinks for this gas in Earth's atmosphere. This is followed by a summary of the contribution that each factor makes to the net carbon dioxide concentration. As you will see:

On the massive scale of the Earth's natural carbon cycles, humans are not in control of current CO_2 levels or the increases in CO_2 concentrations documented during the Industrial Age.

Fossil Fuels as a Source of CO_2

The burning of fossil fuels is currently pumping 33 Gton of CO_2 into our atmosphere every year, with most of the recent increases occurring in developing nations. If *all* of this CO_2 were to remain in the air, atmospheric CO_2 concentrations would increase by 5 ppmv per year. At this rate, CO_2 levels would exceed those seen during the Cretaceous reign of the dinosaurs in less than 2,000 years. Taken in isolation, this statistic sounds alarming. However, two other critical facts must be taken into consideration.

First, the Earth is not endowed with an infinite supply of fossil fuels. Known fossil fuel reserves include[9]: 1) 1.7 trillion barrels of oil, 2) 890 billion tons of coal, and 3) 190 trillion cubic meters of natural gas. The maximum amount of carbon dioxide that can be generated by completely burning *all* of these sources is 5,600 Gton. Assuming that all of this CO_2 stays in the atmosphere (which it doesn't): *The burning of all fossil fuels on the planet would be sufficient to raise the atmospheric CO_2 level by a maximum of 700 ppmv to a grand total of 1,100 ppmv (around three times modern values).* This concentration is insufficient to cause significant global warming (see Chapter 2). This concentration, commonly found in modern greenhouses, poses no threat to either plants or animals (see Chapter 7). *The real concern regarding fossil fuel combustion is that these valuable energy resources will eventually be consumed.* At the current rate of use, the world's fossil fuel supplies will be completely depleted in around 200 years. Unless new technologies are developed (see Chapter 8), our current era of abundant and affordable energy will become a thing of the past.

Second, although humans inject enough CO_2 into the atmosphere to increase concentrations by 5 ppmv/year, the maximum increase reported during the past few years is less than 2 ppmv/year, or only 40% of the emissions total. This discrepancy highlights the fact that fossil fuel combustion is only one of many factors controlling atmospheric CO_2 concentrations.

The Oceans: A Massive Source and Sink for Carbon Dioxide

The vast oceans of Earth exert a major influence on atmospheric carbon dioxide concentrations. The current concentration of dissolved CO_2 and its equilibrium bi-products in seawater (i.e. bicarbonate and carbonate ions, see Appendix 5) of 104 parts per million by weight[10,11] (ppmw) may not seem like much. However, even though only 1/10,000 of the weight of seawater consists of carbon dioxide, consider just how much seawater there is. Oceans cover over 70% of the Earth's surface to an average depth of 3.8 kilometers (or 2.4 miles). The volume of seawater is 1.4 billion cubic kilometers. The mass, or weight, of all of this water is 1.5 billion Gton (1.5 billion billion tons). Because of this:

The carbon dioxide concentration in the oceans of 104 ppmw amounts to 150,000 Gton or 50 times the amount of CO_2 that is present in the atmosphere. This quantity of CO_2 is 30 times greater than the CO_2 equivalent of the Earth's entire fossil fuel reserves, and 5,000 times greater than annual fossil fuel emissions. If all of this CO_2 could be magically released, atmospheric CO_2 concentrations would rise above 20,000 ppmv.

Can the oceans release their massive reserves of carbon dioxide back into the atmosphere? Anyone who has observed CO_2 bubbles forming when they open a bottle of beer or a carbonated beverage knows that the answer to this question is 'yes.' The environmental conditions under which the oceans capture or release CO_2 can be understood by examining simple chemical equilibrium expressions such as Henry's Law[12] (see Appendix 5):

$$[CO_2]_{water} \leftrightarrow [CO_2]_{air},$$

$$K_{Henry} = [CO_2]_{air}/[CO_2]_{water} = 0.033 \text{ M/atm (at } 25°C) \quad \text{(Eq. 4.1)}$$

This expression shows that at equilibrium, the ratio between the amount of CO_2 in air and water in contact with each other is always the same. If CO_2 is added to the air, excess CO_2 will dissolve into the water until the proper ratio is reestablished. Here, since the water is removing CO_2 from the air, it represents a *sink* for the gas. Conversely, if CO_2 is removed from the air, the water will release dissolved gas until equilibrium is reestablished, making the water a *source* for atmospheric CO_2.

Carbonated beverages provide a practical example illustrating Henry's Law. Soda is loaded with CO_2 or is *carbonated* when the liquid is exposed to high-pressure CO_2 gas. The concentration of CO_2 in a typical carbonated beverage[13] is 6,000 ppmw, which is *60 times greater* than the current concentration in our oceans. This illustrates the enormous capacity that water has for adsorbing carbon dioxide. However, when any soda container is opened, gas bubbles form as the CO_2 leaves (making the soda go flat) until the liquid comes back into equilibrium with the much lower CO_2 concentration in our air.

On a planetary scale, if all of the fossil fuels on Earth were instantly incinerated, resulting in an increase in atmospheric CO_2 concentrations from 400 ppmv to 1100 ppmv, the immense volume of our oceans would start to dissolve most of the excess gas until the CO_2 concentration ratio between water and air was eventually reestablished. Even assuming that only the top 10% of the ocean equilibrates with the air (see Temperature Effects below) the new atmospheric CO_2 concentration at equilibrium would be only 480 ppmv. In other words:

Based on known equilibrium constants, the maximum possible permanent increase in atmospheric CO_2 concentrations arising from the combustion of the Earth's entire fossil fuel reserves would be less than 100 ppmv under current climactic conditions.

What happens to the partitioning of CO_2 between the atmosphere and the oceans if climatic conditions change? A second key factor that influences the solubility of CO_2 in water is *temperature*. The Henry's Law constant in Eq. 4.1 is temperature dependent[12] (Fig. 4.2), reflecting the fact that:

CO_2 is less soluble in hot water than it is in cold water.

If you don't believe it, take two identical bottles of soda, one hot and one cold. Open both bottles and see which releases the most gas. If the oceans get warmer, they release CO_2 into the air, while if they get colder they absorb more CO_2 from the atmosphere.

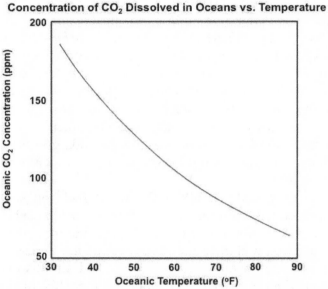

Concentration of CO₂ Dissolved in Oceans vs. Temperature

Fig. 4.2 A curve showing the concentration of carbon dioxide dissolved in ocean water (in ppmw) as a function of temperature. The curve was calculated based on the temperature dependence of the solubility of CO_2 gas in water (Henry's Law, Eq. 4.1) taking into account additional equilibrium constants associated with the acid-base chemistry of CO_2 dissolved in water (see Appendix 5) and assuming that the pH of the ocean is at its current value of 8.2.

The next factor to consider is how temperature and CO_2 are distributed in the ocean. Near the surface, oceanic temperatures range from 28°F (below freezing) in polar waters up to 99°F (37°C) in equatorial waters such as the Persian Gulf. Based on Fig. 4.2, this temperature range corresponds to a range in CO_2 concentrations at the surface by more than a factor of two. The current average near-surface temperature[14] of 63°F (17°C) is used to define 'oceanic temperature' in remaining discussions.

For CO_2 exchange, a more important parameter involves how oceanic temperatures vary with depth. All sunlight is absorbed and converted into heat within a thin (200 meter thick) region called the *photic zone* within which all oceanic photosynthesis occurs. Through turbulence, the photic zone equilibrates with underlying water until it encounters a boundary at a depth of around 1000 meters (1 kilometer) called the *thermocline*.[15] Below the thermocline is the abyssal ocean. For all practical purposes, the abyssal ocean (T = 33°F) does not mix with the overlying water, and never equilibrates with the surface with regard to either temperature or CO_2 concentrations. The near-surface layer engaged in CO_2 exchange constitutes only 20% of the total oceanic volume but still contains on the order of 32,000 Gton of CO_2 (10 times the amount of CO_2 in the air).

The final factor to consider regarding oceanic carbon dioxide is time. When a bottle of soda is opened, it can take hours for the liquid to totally de-gas and come into equilibrium with the air. Now imagine how long it takes for the vast ocean to equilibrate with changes in atmospheric CO_2 concentrations. The slow rate of equilibration represents one of several factors that explain why atmospheric CO_2 concentrations have been rising to a greater extent than expected based on simple equilibrium expressions (see below).

The Vostok Ice Cores

The direct correlation between oceanic temperatures and atmospheric carbon dioxide concentrations is clearly provided by the Vostok Ice Cores[16] (Fig. 4.3). Analyses of these cores provide a detailed historical record of climate over the four most recent major Ice Age cycles into the fifth, including temperatures, snowfall levels, and CO_2 concentrations (from entrapped gas). These ice core samples show that there has been a nearly perfect match between rising and falling CO_2 concentrations (solid dots) and temperatures (continuous curve) over at least the past 400,000 years.

Climate Change advocates would like you to believe that the Vostok Ice Cores support their claim that atmospheric CO_2 levels control the Earth's temperature. However, this claim falls apart when one takes a closer look at the data. First, the Climate Change hypothesis cannot explain why atmospheric CO_2 levels should magically rise *and fall* according to a regular saw-tooth pattern. Such a pattern certainly cannot be attributed to humans or even natural phenomena such as periodic bursts of biological or volcanic activity. Second, a more detailed analysis of the Vostok ice core data show that CO_2 levels increase and decrease *after* the temperature changes, with a time lag of around 1000 years. In other words:

During our current era of recurring Ice Ages, increasing atmospheric CO_2 levels have not caused temperatures to rise. The exact opposite is true. Increasing temperatures have caused atmospheric CO_2 levels to rise.

CO_2 Concentrations and Relative Temperatures in Vostok Ice Cores

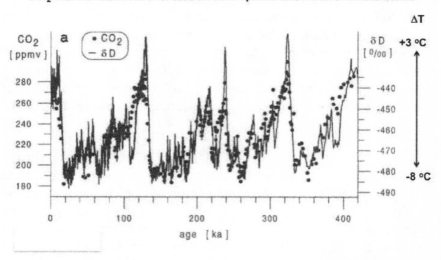

Fig. 4.3 A direct comparison between temperatures and carbon dioxide concentrations during the past four major Ice Age cycles (400,000 years) based on the Vostok Ice Cores[16] (with permission). The solid curve represents the ice's deuterium excess (δD) that is used to calculate ice temperatures. (Extremes in temperature inferred from δD appear on the right. For complete temperature information see Fig. 3.2.) The carbon dioxide concentration (in ppmv) is measured via analyses of entrapped gases and is indicated with the solid dots. Note that the two sets of data appear to be coincident given the time resolution of the graph. Statistical analyses show that changes in CO_2 actually lag changes in temperature by around 1,000 years.

(Incidentally, the time sequencing of the two curves was switched in an Al Gore video aimed at school children.[17] More examples of this sort of behavior are explored in Chapter 9.)

The above observations make perfect sense because: 1) The observed swings in the Earth's temperature are not due to

56

fluctuations in atmospheric CO_2 levels but to fluctuations in the heat reaching the Earth from the Sun. These fluctuations are completely predictable based on periodic oscillations in the Earth's orbit[18] (i.e. the Milankovitch cycles discussed in Chapter 3). 2) Temperature fluctuations at the Earth's surface eventually heat or cool our vast oceans, causing them to either release or absorb CO_2, respectively.

The solubility of carbon dioxide in the oceans confronts the Climate Change narrative with an unsolvable dilemma highlighted by two extreme scenarios:

Scenario #1 – According to Global Warming advocates, the Earth has warmed by 2°F since the start of the Industrial Age. The oceans, which cover 70% of the Earth, dominate the climate. If the climate has increased by 2°F, then oceanic surface temperatures must also have increased by 2°F. If the near-surface ocean has warmed by 2°F, Fig. 4.2 indicates that this temperature change would cause oceanic CO_2 concentrations to drop from 104 ppmw down to 100 ppmw, or by around 4%. This 4% change corresponds to a release into the atmosphere of 650 Gton of the 32,000 Gton contained in near-surface oceanic waters. Such a release would be sufficient to raise atmospheric CO_2 concentrations from 280 ppmv up to 360 ppmv, which is close to the 400 ppmv seen today. In other words:

If oceanic temperatures have really increased as much as claimed, then the increase observed in atmospheric CO_2 levels is due to a warming of the oceans rather than the burning of fossil fuels.

Scenario #2 – According to Climate Change advocates, the entire increase in atmospheric CO_2 levels during the Industrial Age has been due to the burning of fossil fuels. For this scenario to be true, the temperature of the ocean must have been stable or decreasing during that time to prevent oceanic reservoirs from releasing their CO_2 into the air.

If oceanic temperatures aren't actually increasing, there is no global warming. If there is no global warming, then atmospheric CO_2 levels aren't controlling the climate. If atmospheric CO_2 levels *don't control the climate, humans are not causing global warming. Regardless of the scenario, the major conclusion is that humans do not control the climate.*

Scenario #2 comes closest to explaining modern climate changes. Oceanic temperatures have not increased to nearly the extent claimed (see Chapter 6). Instead, humans appear to be injecting CO_2 into the atmosphere faster than the oceans can absorb the excess such that a true state of chemical equilibrium is never achieved. Atmospheric CO_2 levels will continue to creep up until the next natural period of global cooling kicks in to stimulate more rapid absorption of CO_2 by oceanic waters. However, in the meantime, atmospheric CO_2 levels will never become so high as to stimulate measurable global warming.

The Biological Carbon Cycle and the Balance of Nature

The carbon cycle that encompasses the metabolic activity of all living things represents another massive mechanism for adding and subtracting carbon dioxide from the atmosphere. The carbon cycle illustrates the *balance of nature*. Photosynthetic organisms consume CO_2 while generating oxygen (O_2) as a waste product. To counterbalance photosynthesis, Earth's modern ecosystems also contain organisms that consume oxygen as an energy source while replenishing the CO_2 required for continued photosynthesis.

First, consider *sources* of CO_2 in the Earth's biochemical carbon cycle. Breathe in. Breathe out. Every time you exhale, you release CO_2 into the atmosphere. The human body is the equivalent of a biological internal combustion engine, reacting inhaled oxygen with hydrocarbon fuel to produce energy plus CO_2 as a waste product. It is estimated that the average human exhales 1 kilogram (2% of body weight) of CO_2 gas every day.[19] In 'internal combustion' terms, CO_2 emissions per person are the equivalent of the burning of one quart of gasoline per day.

There are currently 7.5 billion humans on our planet.[20] This means that humans exhale 2.7 Gton of CO_2 per year, or around 10% of the current fossil fuel emission total. However, humans represent a single animal species. Cows emit an additional 1.6 Gton, which environmentalists find to be almost as alarming as bovine methane emissions in their calls to ban the consumption of meat and milk by Americans. All livestock in the world contribute 4 Gton/year (14% of the fossil fuel total).[21] Counting people plus all domesticated animals, environmentalists vilify humans for contributing an additional 10 Gton/year (or 40%) to the fossil fuel total. This doesn't count emissions from all the other mammals, birds, reptiles, insects, fish, and other marine creatures that inhabit the Earth. However, over 90% of the CO_2 emissions from all living things arise not from animals, but from anaerobic bacteria and fungi.[22] These organisms metabolize dead plant and animal matter in soil via decay processes that recycle CO_2 back into the atmosphere.

Counting bacteria and fungi, the grand total for the CO_2 produced by all living things is estimated to be 440 Gton/year,[23] or thirteen times the CO_2 currently being produced by fossil fuel emissions. Fossil fuel emissions represent only 8% of biological emissions.

The other half of the biological carbon cycle involves photosynthetic organisms that *remove* CO_2 from the atmosphere. People who are truly concerned about global warming should all plant trees. Every carbon atom in a tree started out as a molecule of CO_2 (see Chapter 7). A single redwood tree contains enough carbon to have consumed all of the carbon dioxide within an area of 40 acres. On land, trees in the rainforest represent 26%, temperate forests 40%, and crops 18% of total carbon fixation. In the oceans, algae remove as much carbon dioxide as plants do over the continents. In total, photosynthetic plants and algae remove on the order of 440 Gton/year from the atmosphere[24] counterbalancing the biological emissions total.

For millions of years, the net carbon cycle has exhibited an almost perfect balance between the rate at which plants and algae fix carbon and the rate at which bacteria and fungi are able to release CO_2 back into the air via the decay of the resulting biomass. How does this carbon cycle respond to external changes in CO_2 concentrations? If CO_2 is added to the atmosphere, plant growth is stimulated to consume this added fuel (see Chapter 7). Additional plant growth leads to the formation of more dead plant matter, which in turn supplies more food to anaerobic bacteria and fungi. At equilibrium, the rates of plant growth and plant decay are once again equal, but both are faster due to an increase in the steady-state CO_2 concentration.

On a temporary basis, the balance between the biological production and consumption of carbon dioxide can be upset. For example, it is estimated that the tragic deforestation of Earth in modern times[25] has eliminated around 30% of its trees, reducing the total amount of carbon fixed by plants by 33%. In the short term, the dead plant matter that built up before the trees were chopped down still represents a food supply for bacteria and fungi. Until this dead matter is consumed - representing three times the current mass of CO_2 in the atmosphere - the net result is a temporary build-up of CO_2.

The 33% decrease in the carbon fixed by plants due to deforestation corresponds to an increase in atmospheric CO_2 of around 75 Gton/year, exceeding annual fossil fuel emissions by a factor of three.

If all photosynthetic organisms became extinct due to some unknown catastrophe, the planetary reserves of dead organic material (1,000 Gton of carbon) would be sufficient to produce 3,700 Gton of CO_2 (roughly twice the current atmospheric total) so long as anaerobic bacteria and fungi remained active.

Geologic Sinks for CO_2

It is important to point out that all changes in carbon dioxide concentrations cannot be attributed to the oceans or the biosphere. The Vostok Ice Core data suggest that the *sum total* of the quantity of CO_2 contained in the air plus the oceans has been more or less constant during the modern era of recurring Ice Ages. This has not been the case over geologic time. Other natural factors must be examined in order to explain the large decreases and increases in atmospheric CO_2 concentrations that occur over millions of years.

Geologic formations represent largely irreversible sinks for CO_2, accounting for the massive amounts of CO_2 that have been removed from the Earth's atmosphere over geologic time (Fig. 4.1). These sinks include fossil fuels and limestone.

Starting with fossil fuels, bacteria and fungi are unable to recycle all dead plant and animal material back into the atmosphere as CO_2. Some of this dead and decaying biomass is eventually buried, compressed, and transformed by geologic processes into the fossil fuels (coal, oil, and natural gas) that we mine today. However, as vast as fossil fuel reserves are, they account for only 5,600 Gton of the CO_2 that has been removed from the atmosphere over geologic time. What happened to the rest?

Most of the carbon dioxide on Earth is currently tied up in geologic formations as carbonate minerals. When CO_2 dissolves in basic oceanic seawater, it reacts with water molecules to form carbonic acid (H_2CO_3) as well as bicarbonate and carbonate anions (HCO_3^- and CO_3^{2-})(see Appendix 5 and Chapter 7). Carbonate anions can be precipitated from water by calcium and magnesium cations (Ca^{2+} and Mg^{2+}) to form carbonate mineral deposits ($CaCO_3$ and $MgCO_3$) called *limestone* and *dolostone*. The White Cliffs of Dover represent just one of these limestone deposits.[26]

The total quantity of limestone deposits on Earth is staggering. With a total mass estimated at 60 billion Gton, limestone represents a quarter percent (1/400) of the entire Earth's crust. *The amount of CO_2 tied up in limestone (25 billion Gton) is over 150 times greater than that found in our oceans, and 10,000 times greater than that found in our current atmosphere.* In fact, it is ten times greater than all of the CO_2 that was present in Earth's primordial atmosphere (not counting the oceans).

The bulk of limestone formation has always been associated with the biological activity of living organisms that extract dissolved carbonates from water to form solid body parts such as seashells (see Chapter 7). Carbonate forming organisms include mollusks, corals, barnacles, tubeworms, and even primitive single-celled creatures such as green algae.

The most extensive limestone formations were created between 2,800 and 600 million years ago by colonies of photosynthetic algae called stromatolites.[27] Stromatolites were also responsible for converting CO_2 into O_2 in our early atmosphere (see the Geologic Record above). This extended period of photosynthetic activity and carbonate formation led to the consumption of vast quantities of carbon dioxide, reducing atmospheric concentrations from 300,000 ppm down to 7,000 ppm.

How significant is limestone formation today? It is important to recognize that the vast quantities of CO_2 that stromatolites consumed were eliminated over a vast number of years. Assuming that most of the CO_2 removal seen during the past 600 million years (Fig. 4.1) is due to carbonate formation, the rate of carbon dioxide removal from the atmosphere has been a more modest 3 million tons or 0.03 Gton/year. At this rate, the CO_2 reserves in our atmosphere and our oceans will last for at least another 250 million years. The bottom line is:

While the irreversible nature of limestone formation makes it the most important sink for CO_2 over geologic time, limestone formation has had a negligible impact on any changes in atmospheric CO_2 concentrations observed during human history.

Volcanic Emissions as a Source of Carbon Dioxide

Volcanic eruptions are significant as they represent the only natural mechanism for injecting new carbon dioxide into the environment. However, environmentalists appear to be conflicted when it comes to deciding whether or not volcanic emissions are important. When attacking fossil fuels, they claim that volcanic eruptions do not emit significant quantities of CO_2 and that volcanoes actually cause global cooling. Conversely, when conservationists in Iceland attack geothermal energy, they claim that CO_2 emissions from geothermal power plants are destroying the planet even though such emissions are negligible when compared with volcanic emissions on their own island. Who is right? To what extent do volcanoes actually contribute to atmospheric CO_2 concentrations?

First, one must understand what volcanoes emit, and how each component of an eruption can influence climate. Emissions having either short or long term effects include volcanic ash, sulfur dioxide, and carbon dioxide. The major atmospheric output of a volcano is ash. Ash injected into the atmosphere blocks sunlight from reaching the Earth's surface and leads to cooling. For example, the eruption of Iceland's Eyjafjoell volcano[28] in 2010 closed airports all across Europe and produced a temporary temperature drop of 0.3°F across the continent. Ash is typically confined to low altitudes where it is washed out by precipitation in less than a year.

Sulfur dioxide emitted by volcanoes gets injected into the upper atmosphere as an aerosol of sulfuric acid droplets. This aerosol also blocks sunlight, like moisture in clouds. However, the volcanic aerosol is not washed out by rain and can persist in the upper atmosphere for more than ten years. Sulfur dioxide is the primary volcanic ingredient responsible for global cooling. For example, the aerosol cloud from the largest eruption in modern history (Mount Tambora in 1815) caused global temperatures to drop by 7°F (4°C) for over a year.

What about global warming? Volcanoes emit two greenhouse gases: water and carbon dioxide. The gas mixture from an 'average' volcano[29] is 85% water, 10% carbon dioxide, and 5% sulfur dioxide. As the atmosphere is already heavily laden with water, the contribution to this gas by volcanoes is negligible. However, the CO_2 emissions could persist after the volcanic ash and sulfuric acid aerosols dissipate to cause warming.

The quantity of CO_2 emitted by volcanoes scales with the size of the eruption. Iceland's Eyjafjoell volcano emitted a paltry 1.5 million tons of CO_2, tying it for 46th place among countries that burn fossil fuels. Moving up in size, the total amount of material spewed forth from Mount St. Helens of 1 cubic kilometer (km^3) was four times larger than the Eyjafjoell eruption. The Mt. Tambora eruption of 1815 ejected 50 km^3. Mega-volcanoes[30] such as Toba (74,000 years ago) and Yellowstone (2.1 million years ago) ejected 2,500-2,800 km^3, representing ecological disasters of the first order. The largest known single eruption ever was the La Gerita event 28 million years ago that ejected a staggering 4,900 km^3 (1,200 cubic miles) of materials. However, even this massive eruption probably only produced around 30 Gton of CO_2, which is about the same amount as current annual fossil fuel emissions.

Single volcanic eruptions are not major CO_2 emitters. What about the sum total of all eruptions on Earth? At this point in time it is estimated that there are on the order of 150 volcanoes degassing CO_2 on land. All of these volcanoes emit approximately 0.3 Gton of CO_2 per year. However, most volcanic activity actually takes place under the ocean along boundaries between tectonic plates.[31] This activity is hidden from view. The most famous of these boundaries is the spreading center called the mid-Atlantic ridge that runs 15,000 kilometers (9,400 miles) from Iceland to Antarctica. No one really knows the extent of undersea emissions. As the oceans cover 70% of the planet, such emissions could be as high as 0.9 Gton/year. However, the U.S. Geological Survey has estimated[32] that total undersea emissions amount to 0.2-0.4 Gton/yr.

Barring a major new eruption, the planetary grand total for volcanic emissions of carbon dioxide is probably on the order of 1 Gton/year, which is much lower that the fossil fuel emission total of 33 Gton/year. Volcanoes have not made major contributions of atmospheric CO_2 concentrations during modern times.

While volcanic emissions are not significant in the short term, over the long term, volcanoes represent the primary source for adding *new* CO_2 to the atmosphere. These additions are typically not associated with single eruptions, but with periods of pronounced volcanic activity extending over thousands or even millions of years. Scientists believe that essentially all of the 300,000 ppmv of CO_2 present in the Earth's primordial atmosphere was produced by volcanic activity. Since then, the most extensive period of volcanism occurred 250 million years ago, coinciding with the largest mass extinction in Earth's history at the Permian-Triassic boundary (see Chapter 7). For a period of up to one million years, repeated eruptions formed a massive lava field within Russia called the Siberian Traps.[33] *The Siberian Traps cover a land area equivalent to the size of modern Europe, with a total volume of three million cubic kilometers (700,000 cubic miles).* This volume is over 1,000 times the size of the largest mega-volcano. The Siberian Trap eruptions introduced on the order of 20,000 Gton of new CO_2 into the atmosphere. When combined with the mass extinction of both carbonate-forming organisms and plants at the end of the Permian (a staggering 95% of all species were driven to extinction), the Siberian Trap eruptions are sufficient to explain the increase in atmospheric CO_2 levels seen between 250 and 150 million years ago.

The Actual Contribution Humans Make to Atmospheric CO_2 Levels

On a geologic time scale of millions of years, the most important source for atmospheric carbon dioxide is volcanoes, while the most important sink for CO_2 involves the biological formation of carbonate minerals. Volcanoes represent the primary mechanism for injecting new carbon dioxide into the atmosphere, while limestone formation represents the main mechanism for irreversibly removing CO_2 from the environment. Most other natural phenomena that impact atmospheric CO_2 concentrations involve changing how carbon and carbon dioxide are partitioned between the atmosphere, the oceans, and life, but do not affect the total amount of carbon that is available. However, while volcanoes and limestone are the only permanent sources and sinks for CO_2, both processes either create or remove carbon dioxide at a rate that is so slow that their effects are only apparent over long time scales. On the timescale of human history, neither process contributes more than 1% to the grand total of CO_2 added to or removed from the air by other sources. Conversely, on a geologic time scale, humans have had no impact (nor will ever have any impact) on atmospheric CO_2 concentrations.

In modern times, the most significant sources and sinks for atmospheric CO_2 include the biological carbon cycle, fossil fuel emissions, and the Earth's oceans. As a source for carbon dioxide, bacteria and fungi are responsible for around 90% of the CO_2 emitted, humans and animal respiration contribute around 5%, while only 5-6% is due to the combustion of fossil fuels. The release of carbon dioxide into the atmosphere is largely compensated for by the photosynthetic activity of plants and algae. When the carbon cycle is in balance, the entire cycle has a negligible impact on atmospheric CO_2 levels. However, cycle imbalances, such as the deforestation of the planet by humans, can lead to significant increases or decreases in CO_2 concentrations. A reduction of photosynthetic activity by 5-6% could create increases in atmospheric CO_2 levels that are equivalent to those that have been observed in modern times.

The oceans represent an enormous reservoir for carbon dioxide, containing 75 times the amount of CO_2 found in the atmosphere. The capacity of this reservoir is largely controlled by temperature. If the temperature stays constant (or drops), the oceans can easily absorb essentially all of the carbon dioxide associated with fossil fuel emissions. However, a temperature increase of only 2°F (1°C) is sufficient to cause the oceans to release enough CO_2 to account for the entire increase in atmospheric levels seen in modern times. Modern oceanic temperatures have been controlled by the Sun rather than by human activities.

Summary: For billions of years, the geologic record conclusively shows that atmospheric carbon dioxide levels do not control Earth's climate. There is absolutely no correlation between the Earth's temperature and CO_2 concentrations. In fact, the worst Ice Age in our planet's history occurred when CO_2 levels were 20 times higher than they are today. Instead of being at an all-time high, CO_2 levels are currently near an all-time low. On the scale of the entire Earth, fossil fuel emissions have little impact in controlling CO_2 levels when compared to natural phenomena including the biological carbon cycle and oceanic temperatures. The Earth's entire fossil fuel reserves could be burned without harming either our climate or life on Earth.

Chapter 5: Ice and Sea Level Changes

Myth: Global Warming is causing all of the ice on Earth to melt, inducing a catastrophic rise in sea levels that will wipe out coastal civilizations and cause ecological disasters such as the extinction of the polar bears.

The most dramatic images used to support the global warming movement involve the constant barrage of videos showing sheets of ice calving off of melting glaciers. These images are always accompanied by voice-overs warning that the Earth is doomed if humans continue to use fossil fuels. In 2008, Al Gore predicted that by 2013, skyrocketing temperatures would eliminate all ice in the Arctic.[1] He also predicted that because of the massive melting of the polar pack ice, sea levels would rise by 20 feet or more, wiping out all coastal cities. Dramatic images in his movie *An Inconvenient Truth* show New York City being inundated by a massive tidal wave caused by global warming.

Has the burning of fossil fuels really pushed the Earth to a "tipping point" beyond which all ice will irreversibly melt? Will Global Melting really lead to the destruction of life as we know it? Below, specific issues associated with planetary ice are addressed. This chapter outlines how much ice there is, how that ice is distributed, to what extent any ice has melted, and the impact that melting ice has had on sea levels and human life. The impact of changes in ice and sea levels on other life forms, including polar bears, is deferred to Chapter 7.

The Current Quantity and Distribution of Ice on Earth

The Area Covered by Ice on Earth

Ice currently covers approximately 15.3 million square kilometers (km^2), 6 million square miles (mi^2), or 10% (1/10) of the land area on Earth.[2] The land covered is slightly less than the area of South America. Sea or pack ice at the North and South Poles covers an average of 21.5 million km² (8.2 million mi²) or 6% of the ocean's surface.³ This area is similar to that of North America. At the North Pole, this pack ice fills most of the surrounding Arctic Ocean, while at the South Pole pack ice encompasses much of the surrounding Antarctic continent. The total area covered by ice on both land and sea is around 36.8 million km² (7% of the Earth) or an area equivalent to that of Asia.

The Volume or Mass of Ice on Earth

The most important measure of the amount of ice present is given by its volume or mass rather than by its surface area. The volume of ice is equal to its area times its thickness, and is represented in units of cubic kilometers (km³) or cubic miles (mi³). The weight of the ice is calculated by multiplying its volume times its density of 0.92 grams per cubic centimeter (assuming that the ice is solid rather than porous). The United States Geologic Survey (USGS) estimates that the total amount of ice on Earth is currently 30.5 million km³ weighing 28 million Gton (where a Gton is one billion tons).⁴ This is an enormous quantity, yet it represents only 1.7% of the total liquid and solid water on Earth.

Ice Melting and Its Impact on Sea Levels

The most important quantity in most global warming arguments is the volume of water that would be produced if the ice were to melt. This water volume determines how much sea levels could potentially rise. Because ice has only 92% of the density of water, the melting of all ice on Earth would generate 28 million km³ (6.7 million mi³) of water.

*If **all** ice on Earth were to melt, the amount of water released would cause sea levels to rise 67 meters (73 yards) or three quarters the length of a football field.*

The Distribution of Ice and Potential Impact on Sea Levels

Antarctica and Greenland combined represent 99% of all ice on Earth. The remaining 1% is distributed between temperate glaciers, ice sheets, and pack ice. Below is a list of the major categories of ice on Earth in order of decreasing importance. For each category, the area, volume, and maximum sea level rise (see Appendix 6) that would accompany the total melting of each source are tabulated.[2]

The Antarctic Ice Cap: Almost all ice on Earth (87%) is found over the continent of Antarctica. The massive Antarctic Ice Cap covers almost the entire continent and can exceed three miles (5 kilometers) in thickness.
Area: 13 million km^2 (5 million mi^2) or 85% of all ice covered land
Volume: 26.5 million km^3 (6.3 million mi^3) or 87% of all ice
Maximum Sea Level Rise Upon Melting: 58 meters or 63 yards

The Greenland Ice Cap: The second largest accumulation (12%) of ice covers the large island of Greenland to an average depth of almost two kilometers or 1.2 miles.
Area: 2 million km^2 (760,000 mi^3) or 13% of ice covered land
Volume: 3.7 million km^3, 880,000 mi^3 or 12% of all ice on Earth
Maximum Sea Level Rise: 8 meters (9 yards)

Temperate Glaciers and Ice Caps: More humans live near temperate, high altitude glaciers than any other major ice form. The advances and retreats of temperate glaciers have been monitored throughout history. However, such glaciers represent less than 1% (1/100) of the ice on Earth.
Area: 725,000 km^2 (275,000 mi^2) or 5% of ice covered land
Volume: 300,000 km^3 (71,000 mi^3) or 0.9% of all ice on Earth
Maximum Sea Level Rise: 70 centimeters (27 inches)

Shelf Ice: Shelf ice originates over the Antarctic continent. Shelf ice extends out over the ocean as extensive sheets that can be over 200 meters (or yards) thick. Massive icebergs that tower hundreds of feet above the ocean originate either from shelf ice or continental glaciers. However, shelf ice represents less than half a percent (1/200) of the ice on Earth.
Area: 910,000 km^2 (345,000 mi^2)
Volume: 160,000 km^3 (38,000 mi^3) or 0.5% of all ice on Earth
Maximum Sea Level Rise: 35 centimeters (14 inches)

Pack Ice: Although pack ice covers almost 6% of the entire ocean,[3] the average thickness of this ice is only around 2 meters (6 feet). Because it is so thin, the total volume of sea ice at both poles is less than 1/700 or 0.14% of the total amount of ice on Earth.
Area: 21.5 million km^2 or 8.2 million mi^2
Volume: 44,000 km^3 (10,000 mi^3) or 0.14% (1/700) of all ice
Maximum Sea Level Rise: 10 centimeters (4 inches)

It is important to note that the maximum sea level rises listed for pack and shelf ice are based on the volume of water that would be produced rather than the impact of that water on sea levels. This is because if ice is floating in water rather than resting on land, the melting of that ice causes no increase in the level of the surrounding water (see Appendix 6). *The actual sea level rise for the melting of all pack ice and essentially all shelf ice is zero.*

Below, the extent of melting associated with each of the above ice reservoirs is examined. However, keep in mind the total quantities of ice available within each reservoir when evaluating the potential impact of global warming. For example, as incredible as it may seem, all pack ice on Earth could completely melt without causing sea levels to rise at all. All temperate zone glaciers could completely melt without causing sea levels to rise more than two feet. The only way to create truly significant increases in sea level is to melt the massive ice caps in both Antarctica and Greenland. To what extent can this happen?

Historical Ice and Sea Water Levels

Modern variations in ice and sea levels are negligible when compared with those that accompany natural Ice Age cycles.

The geologic record provides a perspective on how climate impacts the quantity of ice on Earth. Over geologic time ice levels on Earth have encompassed every extreme. During the Cryogenian Period (800 million years ago), the planet was almost entirely encased in ice.[5] Since then, there have been many extended periods when essentially no ice was present (see Chapter 3). As recently as 3 million years ago, sea levels were 165 feet (50 meters) higher than now, indicating that ice levels were only 25% of modern values. Even during the most recent era of recurring Ice Ages, there has often been less ice than there is today. For example, only 125,000 years ago (during the Stone Age), water levels were 18 feet (5.5 meters) higher than today[6] (or approximately the sea level rise predicted by Al Gore). In contrast, ice covered almost 1/3 of the entire planet during the last Ice Age, resulting in sea levels that were 400 feet (122 meters) lower. Such low water levels allowed ancient peoples to cross the Siberian Land Bridge to populate North America. The changes that have taken place in the average sea level since the last Glacial Maximum are shown in Fig. 5.1.[7]

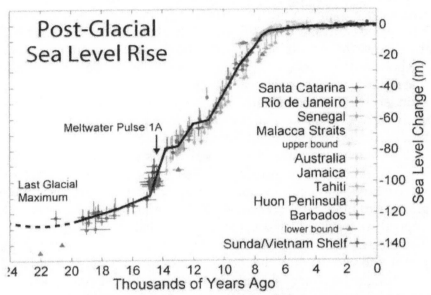

Fig. 5.1 Geologic records of sea level changes over the past 22,000 years. Sea levels have risen by over 130 meters (120 yards) since the height of the last Ice Age. Sea levels have risen by around 5 meters during the past 7,000 years. (From Ref. 7 with permission.)

Most of the ice melted between 19,000 and 7,000 years ago. While ice has continued to melt during the past 7,000 years, the total amount of ice since then has been remarkably stable, increasing sea levels at a rate of a few centimeters (or inches) per century.

The extent to which each component of the total ice inventory has been impacted during the Industrial Age is outlined below. The discussion progresses from those components that receive the most media attention to those that are rarely discussed. Ironically, this progression also moves from those components that are the least important to those that are truly significant.

Modern Variations in Pack Ice

Around 80% of pack ice melts and reforms with the seasons every single year. The average extent of pack ice has only decreased by a few percent in the Industrial Age.

In 2007, Al Gore predicted that the Arctic Ocean would be completely ice free by 2013, leading to the extinction of many Arctic species including polar bears.[1] Climate Change advocates have since retreated from this statement because it obviously isn't true. However, paralleling their reports that each successive year is the hottest on record (see Chapters 6 and 9), they continue to report that the extent of the pack ice at both the North and South Poles has been shrinking dramatically each and every year. Specifically, advocates point to the summer of 2012, during which ice levels in the Arctic Ocean reached all-time lows (see below).

Emboldened by the 2012 reports, Australian Professor Chris Turney launched an expedition in December of 2013 to prove that the Antarctic pack ice was undergoing catastrophic melting due to global warming. However, much to his surprise, his ship was soon trapped in sea ice that became so thick that the vessel could not even be rescued using modern icebreakers.[8] Professor Turney was ruefully forced to admit that he got "stuck in our own experiment."

Professor Turney should have known that a more accurate means of determining the size of the pack ice involves the use of satellite imaging. Satellites have been taking pictures of the poles every day since 1981. The National Snow and Ice Data Center (NSIDC) is one many organizations that is constantly analyzing satellite images of pack ice. Fig. 5.2 contains some of their observations regarding the amount of the pack ice present in both the Arctic and Antarctic Oceans.[3] Note that even satellite images are subject to interpretation because: 1) the images show the *area* covered by ice, but not its *thickness* or *volume*, and 2) the pack ice is not a continuous sheet of ice. The area covered by ice often contains isolated ice floes as well as areas of open water. Each investigator defines an arbitrary boundary for the pack ice based on a percentage of water covered. In Fig. 5.2, this boundary includes that area of the ocean containing at least 15% sea ice.

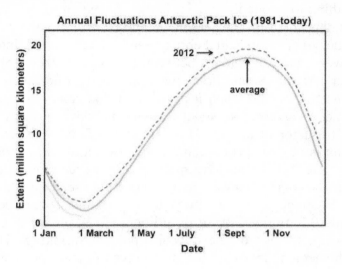

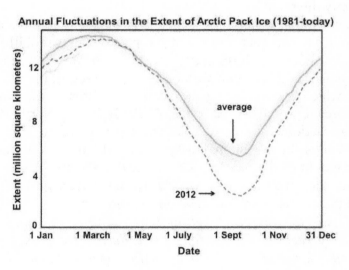

Fig. 5.2 The area covered by pack ice as determined from satellite imaging of both the North (top) and South (bottom) Poles in millions of square kilometers. Each curve shows the fluctuations in pack ice that occur over the course of a year. The solid curve represents the average over all years since 1981, the grey band represents the normal spread between years, and the dashed curve represents the anomalous year of 2012. (Adapted from Ref. 3.)

In other words, in their analyses, there is substantially less ice contained within the sea ice boundary than there would be for a continuous solid sheet of ice.

The first thing to notice in Fig. 5.2 is that the extent of the pack ice is highly dependent on the season. At the North Pole, the pack ice covers the greatest area in February and March after winter, while the covered area is smallest between September and October after summer. At the South Pole, the opposite is true, as it is summer at the South Pole when it is winter at the North Pole. At the North Pole, the average seasonal variation is between 6 and 15 million km^2 (a factor of around 3), while at the South Pole even more variation is observed (between 2.5 and 18.5 million km^2 or a factor of 7). Note that the massive melting of pack ice that occurs each and every year is not causing mass extinctions (see Chapter 7). The average coverage over an entire year is nearly the same at both poles (around 10.5 million km^2 or 4 million mi^2). While ice at one pole is growing, the ice at the opposite pole is shrinking. This means that the net amount of sea ice at any given time of the year is almost exactly the same.

The second thing to notice in Fig. 5.2 is that sea ice coverage can vary from year to year at each of the poles. This variability is not identical at both the North and South Poles, or for each season in a given year. The maximum variability is observed in the degree of melting which occurs during the summer months at the North Pole. The area covered by pack ice in April at the North Pole fluctuates at random by around 2%. The dashed line in Fig. 5.2 shows the results for the anomalous year of 2012 that set a record low for the extent of the Arctic pack ice for the September season. In contrast, the pack ice in Antarctica reached record highs throughout that entire year. A critical observation is:

Regardless of the degree of melting in the summer, the pack ice has completely recovered its original size during the subsequent winter months for every year since satellite measurements have been taken.

The third thing to notice is that sea ice has been stubbornly resisting Al Gore's predictions. The pack ice is still there. In fact, the annual average coverage by sea ice is essentially the same as it has been since satellite observations commenced in 1981. However, this has not stopped advocates, including those at supposedly reputable government agencies including the National Oceanic and Atmospheric Administration (NOAA), from cherry-picking the data in blatant attempts to mislead the public (see Chapter 9). One ploy used by NSIDC (illustrated in Fig. 5.2) is to use their website to display results that highlight melting, such as the data for the North Pole in 2012.

Another ploy is to select only that location and time of year (September at the North Pole) showing the greatest degree of variability and melting. However, even here, the evidence for global warming is less than compelling. As shown in Fig. 5.3, over the past ten years - while annual CO_2 emissions have almost tripled - the minimum extent of sea ice at the North Pole is currently less than 3% lower than the ten-year average.

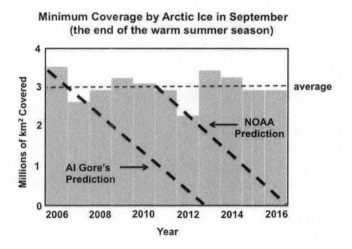

Fig. 5.3 A bar graph showing how the coverage of pack ice, which is at a minimum in September, has varied between 2006 and 2016. The results do not show a significant decrease in recent history. Dashed lines show how both Al Gore and NOAA have used results from two years in succession to make alarmist long term predictions regarding the fate of the pack ice. (From data in Ref. 3.)

The final ploy aimed at convincing the public that global warming is real is to only provide reports on those years that support the warming hypothesis (see Fig. 5.3). There was a substantial drop in the minimum sea ice coverage at the North Pole

75

between 2006 and 2007, which led to Al Gore's prediction that all sea ice would be gone by 2013. There was another more extended drop between 2010 and 2012 that led gleeful climate change advocates at NASA to pronounce that this melting was the 'canary in the coal mine' that proves that global warming is destroying the planet.[9] However, these same NASA scientists were mystified by what happened in 2013 and 2014, when the minimum in sea ice coverage in the Arctic *grew* by 1.7 million km^2 (twice the size of Alaska).[10] The minimum sea ice coverage near Antarctica also increased. Did this mean that Earth was headed for another Ice Age? Perhaps. One NASA scientist went so far as to say that the massive growth of the Antarctic ice sheet was due to 'excess snowfall produced by global warming.' If so, had the trend not reversed itself, it would have taken only a few more years before catastrophic global warming converted the entire Earth into a frozen ice ball.

*If there has been <u>any</u> decrease in the fraction of the ocean covered by sea ice, it has been less than a percent since continuous satellite observations started in 1981. Even if **all** of the pack ice on Earth were to melt, there would be no rise in sea levels at all.*

Changes in Antarctic Shelf Ice

Antarctic shelf ice is largely intact.
Melting of the shelf ice has had a negligible impact on sea levels.

A climate catastrophe that environmentalists always point to is the destruction of Antarctica's shelf ice formations. The two major ice shelves in Antarctica are the Ross Sea Shelf and the Ronne Sea Shelf.[11] The Ross Ice Shelf covers an area of 487,000 km^2 (larger than the state of California), is around 200 meters thick, and has a total volume of 97,000 km^3. The Ronne Ice Shelf has an area of 422,000 km^2, a thickness of around 150 meters, and a total volume of 63,000 km^3. The two Antarctic ice shelves combined have a total volume of 160,000 km^3, or about half as much as all non-polar glaciers and ice caps combined. However, because the ice shelves are floating in the ocean, the melting of all shelf ice would result in zero increase in planetary sea levels (see Appendix 6).

Progressives claim that global warming is destroying the ice shelves around Antarctica by two distinct mechanisms. The first mechanism involves simple melting, where warming oceanic currents are eroding the ice sheets from underneath. It has been reported that the Ronne Ice Shelf is melting from below at the rate of 28 cm (11") per year.[12] This corresponds to 113 Gton, but would not cause any rise in sea levels (see Appendix 6). However, a bore-hole drilled through the shelf ice in west Antarctica in January of 2010 casts doubt on even this low number, as researchers found that the water under the ice has not warmed at all and is not warm enough to cause significant melting.[13]

The second mechanism involves the calving of massive icebergs. Cracks can form in the ice shelves. Depending on where the cracks are and how deep they are, they can release enormous icebergs into the ocean. Advocates claim that global warming has created new cracks that will cause all of the ice shelves to float away. They have no basis to support this claim. The largest iceberg ever reported broke off of the Ross Ice Shelf in March of 2000.[14] Iceberg B-15 had the incredible area of 31,000 km^2 (larger than the state of Vermont). Had it been on land, this single object contained enough water to have raised sea levels by 1.4 cm. However, as this massive iceberg was already floating, no rise in sea levels was produced (see Appendix 6).

Changes in the Extent of Temperate Glaciers

Temperature glaciers have been receding during much of the Industrial Age. The rate of glacial retreat has caused a measurable increase in planetary sea levels.

Progressives have been warning that most of the glaciers on Earth are melting so rapidly that they will be completely gone within a generation. In 1999, the Intergovernmental Panel on Climate Change (IPCC) warned that all glaciers in the Himalayas would be completely gone by 2035.[15] Al Gore predicted that there would no longer be any ice on Africa's famous Mount Kilimanjaro by 2015.[16] However, the IPCC was ruefully forced to admit that the first prediction was lifted from a telephone interview with obscure Indian scientist Syed Hasnian reported in the magazine *New*

Scientist in 1999.[17] It has no basis in fact. The second prediction has also been proven wrong. There is still a substantial ice cap on Mount Kilimanjaro that scientists predict will still be present at least fifty years from now.[18] In addition, it is well documented that 50% of the retreat of Kilimanjaro Glacier occurred in the 56 years between 1880 and 1936 (before the major fossil fuel era), whereas it has only retreated by an additional 30% in the 80 years since.

Satellite imaging is currently being used to continuously map Earth's glaciated regions.[19] These images show that the planet contains over 200,000 glaciers. Glaciers cover 725,000 km^2 (275,000 mi^2) or 5% of the land area on Earth. Much of this land is above the Arctic Circle or below the Antarctic Circle, feeding from the massive ice sheets in Greenland and Antarctica (discussed below), as well as territory in Canada, Alaska, and Siberia. The glaciers discussed here reside in mountainous terrain in more temperate climactic zones. It is estimated that the total volume of all of these glaciers is 300,000 km^3 (70,000 mi^3).

Each glacier exhibits its own unique size, shape, thickness, volume, and behavior. Some glaciers are currently receding, while others are either stationary or advancing.[20] Advancing glaciers include Hubbard Glacier (the largest glacier in North America) and the Perito Moreno Glacier (the largest glacier in South America). Retreating glaciers include the largest glacier in Asia (the Siachen Glacier in the Himalayas), as well as the largest one in Europe (Iceland's Vatnajokull Glacier). Over the entire Earth, receding glaciers outnumber those that are advancing. For example, within the massive (40,775 km^2) ice field spanning the border between India and Pakistan in the Himalaya-Karakoram region, 35% of the glaciers are advancing, while 65% are retreating.

How can it be that some glaciers are advancing while others are retreating even within the same geographic area? Researchers have determined that since the end of the Little Ice Age in 1850, temperature is not always the primary climactic factor controlling a glacier's fate. One obvious factor is whether or not snow falls on a given glacier faster than the resulting ice can melt (see Chapter 3 and Greenland below). Another important factor involves whether the air above the glacier is dry or humid. For example, while the glacier atop Mount Kilimanjaro has been abating since 1880, much of the retreat is not due to melting. Central Africa has actually been cooling for the past 30 years. Since 1979, temperatures at the summit have averaged -7°C. Even the maximum reported temperature of -1.6°C is insufficient to melt ice. However, the air above Kilimanjaro glacier is dry enough that ice can evaporate directly into the air by a process called sublimation. It has been shown that the air above Kilimanjaro has become dryer due to local deforestation,[16] leading to the glacial retreat.

Although the behavior of each glacier is unique, the 'average' temperate glacier is indeed retreating. One of the most documented retreats is that of the Vatnajokull Glacier in Iceland. Vatnajokull is the second largest temperate-zone glacier, with a total volume of 3,100 km³ (1% of the total temperate glacial volume). This glacier has advanced and retreated many times during human history (see Fig. 3.5). Extensive studies indicate that this glacier is currently receding at a rate of 10 km³/year. If this rate continues, the glacier will disappear in around 300 years.

Unfortunately, it has been difficult to do more than obtain rough estimates regarding the volume of water being released by most glaciers because: 1) satellite images only monitor the *area* covered by the glacier, but not its *thickness or volume*, and 2) many glaciers are in remote regions where glacial thicknesses have not been determined. Estimates for the volume of ice present in the Himalaya-Karakoram region range from 3000 to 4800 km³. As a result: *Estimates for the sea level rise associated with the melting of temperate glacial over the past 100 years span a large range from 3 to 8 centimeters.* (IPCC obviously reports the maximum.) This means that the 'average' glacier can expect to survive somewhere between another 1,000 to 2,000 years. Over 100 years, the melting of temperate glaciers could cause sea levels to rise as much as 8 centimeters or three inches (see Sea Levels below).

Recent Changes in the Massive Greenland Ice Cap

The Greenland Ice Cap has been retreating from the perimeter, but is thickening in the interior. The impact of Greenland on rising sea levels is uncertain, as the ice cap may be in a transition between shrinking and growing.

The Greenland Ice Cap is the second largest accumulation of ice on Earth. It has retreated and advanced multiple times during human history. Greenland is an island territory that until recently was administered by Denmark. This is because Danish Vikings colonized the island[23] in 986 A.D. during the Medieval Warm Period and established several thriving farming communities along the coast (Chapter 3). Unfortunately, the Vikings were forced to abandon Greenland by 1500 A.D. in the face of advancing glaciers during the Little Ice Age (Chapter 3). The end of the Little Ice Age in 1850 triggered a general glacial retreat that commenced long before humans started introducing significant concentration of carbon dioxide into the atmosphere. There is still more ice in Greenland than there was when the Vikings first colonized it.

Climate Change advocates claim that the Greenland Ice Cap is currently melting at an alarming rate. As evidence, they compare satellite images taken in August of 2011 and 2012 to show that the glaciers of Greenland have been retreating from the coastline.[24] By analyzing these images, they claim that Greenland is losing ice at the rate of 300 Gton per year. Unfortunately for their argument, these images provide information regarding the ice cap's area but not its thickness. Thickness data are required in order to calculate changes in the ice cap's volume.

The Danish Meteorological Institute has been monitoring the thickness of the ice cap since at least 1990. Their findings[25] (Fig. 5.4) show that starting each year in October, snowfall has been accumulating on the ice cap fast enough to cause its total mass to increase. This increase continues until the end of May, by which time around 550 Gton of new ice (the equivalent of around a foot of snow over the island) has been deposited. During the following three summer months, warmer temperatures start to induce melting. As the melt rate overtakes the snow deposition rate, the total weight of the ice cap starts to decrease. The net annual growth of the ice cap corresponds to where each annual curve intercepts the graph at the end of August, after which a new yearly cycle starts. The results show that over the last 16 years:

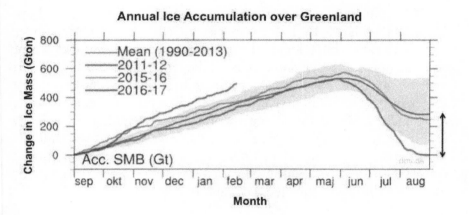

Annual Ice Accumulation over Greenland

Fig. 5.4 Monthly measures of the surface mass balance (SMB) in billion tons (Gt) of ice that has accumulated over Greenland since 1990. This mass is proportional to the total ice volume. Every year, ice accumulates between September and May. Melting exceeds accumulation between June and August. Solid curves are for specific years. The bottom curve is for the anomalous year of 2012. The double black arrow to the right indicates the average annual snow accumulation of around 300 Gt. (Adapted from Ref. 25).

The Greenland Ice Cap has been getting thicker rather than thinner, and is growing rather than shrinking in terms of its total ice volume. The Danes estimate that the average growth rate for the Greenland Icecap over the past 16 years has been 300 Gton/year.

If the current trend continues, it is anticipated that the increased ice mass in the interior will eventually work its way to the coast under the influence of gravity, causing the glaciers around the exterior of the island to advance rather than retreat.

Note how Climate Change advocates have skewed the Greenland results in an attempt to prove that the ice cap is melting due to global warming: 1) Their findings are based on satellite images showing signs that the outer boundary of the ice sheet is melting, while ignoring the vast interior where ice has been accumulating. In fact, if one ignores accumulation, the melting from June to August reported in the Danish data (Fig. 5.4) is actually in good agreement with the 300 Gton losses estimated from satellite images alone. 2) Just as they have done in reporting the supposed disappearance of the Arctic pack ice, their conclusions focus on the specific fall season in the anomalous year of 2012 which saw the maximum amount of melting in both Greenland (Fig. 5.4) and the pack ice (Fig. 5.2). Over longer periods of time, much less melting has been seen, and in fact recent observations show that some coastal glaciers have either not moved or have actually advanced. 3) Their highest (and erroneous) claim for melting of 300 Gton/year may seem like an enormous number, but on the scale of the Greenland Ice Cap (weighing 3.4 *million* Gton), it is negligible. The Greenland Ice Cap will not be disappearing any time soon. Even assuming the maximum current rate of melting, it will still be there ten thousand years from now or long after the next major Ice Age has commenced.

Recent Changes in the Dominant Ice Repository on Earth: The Antarctic Ice Cap

Over 99% of all ice in the enormous Antarctic ice cap is far too cold to melt. Glaciers are currently retreating along the west coast of the Antarctic Peninsula, impacting less than half a percent (1/200) of all Antarctic ice. Snow and ice are accumulating on the rest of the continent. The net impact of Antarctic ice on modern changes in sea levels is uncertain.

The massive Antarctic Ice Sheet contains 87% of all ice on Earth. The global warming community and the media are constantly warning that Antarctica is warming at an alarming rate, and that its massive ice reserves could melt any day now to create untold ecological disasters.

Before continuing, one fact needs to be made perfectly clear: *The Antarctic continent is by far the coldest place on Earth.*

The continent is so vast that it encompasses 5 climate zones, each one of which is frigid.[26] The largest zone is the high plateau region, where most of the ice resides. Here, monthly average temperatures range from a *high* of -30°C (-22°F) down to a low of -60°C (-76°F). The freezing point of water is 0°C (32°F). This means that:

The extent of global warming that would be required to start melting the bulk of the Antarctic Ice Cap is a temperature increase of over 54°F (30°C).

In the low plateau and the high latitude coastal regions, temperatures are also insufficient to melt ice, with highs in the summer months reaching -12°C (10°F) and -2°C (28°F) in these two respective regions. The 'torrid tropics' of Antarctica include the low latitude coastal area (east of the Ross Ice sheet) and the Antarctic Peninsula (along the west coast). Here the temperature can actually creep up to +2°C and +1°C, respectively for a few weeks during the summer (just above the freezing point of water). However, even in these zones, the temperature is -30°C (-22°F) and -15°C (5°F), respectively, for almost the entire year (Fig. 5.5).

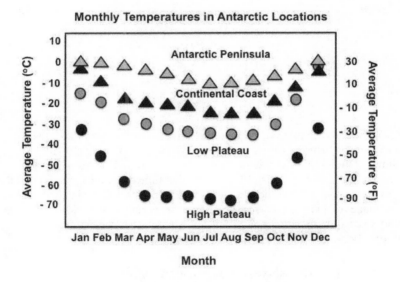

Fig. 5.5 Average monthly temperatures (in °C) reported for various ground stations around Antarctica. The dark circles represent the continental high plateau where most of the ice on Earth resides. The warmest region is in the Antarctic Peninsula indicated by the gray triangles. (Adapted from Ref. 26)

Regardless of land-based temperature records, environmentalists claim that the Antarctic continent is warming and melting at a catastrophic rate. Is this true? First, satellite measurements taken since 1982 indicate the extent to which temperatures over the continent have changed in recent years.[27] These measurements show (Fig. 5.6) that most of the continent has been cooling rather than warming.

Changes in Antarctic Temperatures (1982 - 2004)

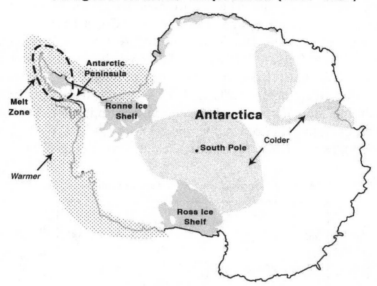

Fig. 5.6 Trends in Antarctic temperatures (in °C/year) from 1982-2004 as collected from sensors flown on National Oceanic and Atmospheric Administration (NOAA) satellites (adapted from Ref. 27). Areas labeled 'warmer' increased by 0.1-0.2 °C/yr over that time span, those cooling by 0-0.1 °C/yr are in white, while those that cooled by up to 0.2°C are labeled 'colder.' Note that even in those areas showing increasing temperatures, temperatures rarely exceed the freezing point of water. Most of the glaciers that are disappearing are in the 'melt zone.'

The only area showing significant warming is near the west coast of the Antarctic Peninsula. In the small region showing the maximum warming, coastal areas can now experience temperatures exceeding the freezing point of ice for up to three months during the summer. In addition, the northern tip of the peninsula is by far the wettest part of Antarctica, receiving 14-20 inches of precipitation per year. In the summer, this precipitation sometimes falls as rain to assist in the melting process.

The Antarctic Peninsula does not represent the entire Antarctic continent. The peninsula has a land area of 522,000 km² (4% of the continent) and contains less than 0.4% (1/250th) of all Antarctic ice. If melted, *all* of this ice would raise sea levels by around 9 inches (23 centimeters). However, even accounting for recent temperature increases, the east coast of the peninsula, as well as the central highlands, almost never experience sufficient heating to melt ice. The only region undergoing melting involves glaciers entering the sea along parts of the west coast that see the highest temperatures. The melting of all ice within the effected zone would raise sea levels by less than three inches. As with most glaciers, there is a high degree of uncertainty regarding current melt rates, which are estimated to be 0.22±0.16 mm/year.[27] Based on this broad estimate, the west coast glaciers will be completely gone in anywhere from 20 to 130 years.

What about the rest of the vast Antarctic continent? The media is curiously silent about most of Antarctica, as events on the continent do not support their agenda. As in Greenland, snow and ice have been accumulating away from the western coastline. In fact, NASA scientists have reported[28] that the Antarctic ice sheet showed a net gain (i.e. continental accumulation minus the melting of west coast glaciers) of 112 Gton per year between 1992 and 2001 and 82 Gton/yr between 2003 and 2008. *This means that Antarctica has probably been contributing to a lowering rather than a raising of sea levels* (see below).

Modern Sea Level Changes: Claims versus Facts

Claims for Rising Sea Levels

Based on information compiled above regarding the fate of each major source of ice on Earth, one can predict how much sea levels should rise as a result of melting ice, and compare such predictions with measured sea level changes. However, before doing so, it is useful to review some of the predictions that are striking fear in the hearts of those who listen to the media and the Global Warming movement. Premises include that: 1) carbon dioxide controls the temperature of the Earth (not true as shown in Chapters 2-4), 2) humans control carbon dioxide levels by burning fossil fuels (see Chapter 4), 3) emissions are increasing at an exponential rate (see Fig. 1.1), so therefore 4) the temperature of the Earth (see Chapter

6), ice melting, and sea level rises must also be increasing at an exponential rate.

Al Gore's Nobel Prize winning movie *An Inconvenient Truth* provides multiple images of the consequences of rising sea levels on humanity if the use of fossil fuels continues.[29] He has predicted that global warming will cause sea levels to rise by twenty feet or more. He shows Florida being completely submerged, while coastal cities such as San Francisco are being annihilated. He claims that millions of people will be flooded out, stranded, and even killed by the rising waters, including tens of millions near Beijing, 40 million near Shanghai, and 50 million in Calcutta and Bangladesh.

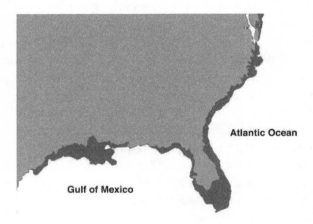

Fig. 5.7 A map of the Gulf Coast and East Coast of the United States showing dark regions that would be underwater if sea levels were to rise by ten meters (33 feet).[31] Two things to note are: 1) even the extreme rise of ten meters would be insufficient to flood most of the United States, and 2) at the current rate of melting, it will take 6,700 years to produce a rise of ten meters, which a) will give humans ample time to move to higher ground, and b) is well beyond the time when the next major Ice Age (and precipitous drops in sea level) will occur.

The Union of Concerned Scientists has predicted that extensive melting in Greenland and west Antarctica will soon cause sea levels to rise between 12 and 40 feet.[13] The IPCC has increased their 2007 projections of a sea level rise of 18 and 59 cm (7 inches to two feet) by the year 2100 up to 1.2 meters (4 feet) in their 2013 Climate Change Report.[30] In 2010, the United Nations claimed that increasing combustion of fossil fuels will catalyze even more rapid melting, causing sea levels to rise by as much as 270 cm (9 feet) by

2100.[13] Not to be outdone, the United States Geologic Survey has prepared a 'fact sheet'[31] predicting a sea level rise of ten meters (33 feet) that they claim will flood 25% of the U.S. population accompanied by a map showing the regions of the Gulf Coast and East Coast that will soon be under water (Fig. 5.7). Should people really be worried about such an ice-melt apocalypse?

Summary of Ice Melt Estimates

What claims regarding sea level rises fall within the realm of possibility based on ice melt estimates? Table 5.1 represents a quick summary of all ice melt information provided in this chapter.

Table 5.1

Potential Contributions to Modern Sea Levels Due to Ice Melt

Source	All Ice (cm)	% Melted	Rise Based on Melt Claims (cm/100 yr)	
			Max.	Min.
Pack Ice	10	3%	0.3	0
Ice Shelves	35	0.7%	0.25	0
Glaciers	70	4 to 11%	8	3
Greenland	800	-1 to +1%	8	-8
Antarctica	5800	-0.05 to 0.7%	41	-3
Total Ice	6715	0.1 to 0.8%	57.5	-8

Notes: One inch = 2.54 centimeters. The actual sea level rise for pack and shelf ice is actually close to zero (see Appendix 6).

Excluding the most outlandish claims, Table 5.1 contains a summary of how much melting has been estimated for each major ice source on Earth, and the extent to which such melting should contribute to rising sea levels. Table 5.1 is organized according to

87

the amount of ice involved per source. The first column represents the sea level equivalents in centimeters corresponding to the total quantity of ice that is available. The second column lists the amount of melting that is claimed for each source per 100 years. Note that there are large uncertainties associated with most entries. The third and fourth columns list the sea level rise that would result from the maximum and minimum degree of melting based on estimates in the second column. The media and climate change advocates invariably report the maximum entries listed in the third column.

The maximum contribution to modern sea level changes due to the melting of both pack ice and Antarctic shelf ice is close to zero (see Appendix 6). In contrast, the melting of temperate glaciers is clearly substantial, resulting in anywhere from 3 cm to 8 cm per hundred years of the observed rise in sea levels. The ice masses of both Antarctica and Greenland are so massive that the melting of even a tiny fraction of their total volume would represent a substantial contribution to sea level changes. Unfortunately, experimental uncertainties are high regarding even those measurements taken over the past 15 to 25 years. Depending on whose measurements are to be believed, Greenland could be either adding or withdrawing up to 7 cm of water from the oceans per 100 years. While glaciers along the west coast of Antarctica are clearly melting, ice on the remainder of the continent is accumulating. Estimates for the resulting sea level change range from a drop by 3 cm to an increase of 41 cm per hundred years. Adding up the maximum and minimum values for each ice source, sea levels could have changed anywhere between a drop of 8 cm (3") to an increase of 54 cm (22") during the past hundred years. How do the values reported for ice melt compare with actual sea level rises as well as the claims of the global warming community?

Actual Sea Level Rises

In 2013, the IPCC projected that global warming will cause sea levels to rise up to 1.2 meters or 4 feet over the next hundred years. It looks like the IPCC wasn't even examining its own compilations of sea level data reported since 1880 (Fig. 5.8).[32]

Over the past 136 years, even the IPCC compilation shows that sea levels have risen by a maximum of 20 centimeters, leading to a calculated sea level increase of around 15 centimeters (6 inches) per hundred years. Since around 1910, *this rise has been constant at a slow but steady linear rate. In contrast, fossil fuel emissions were low from 1880 to 1960, but increased by a factor of ten between 1960 and today. Projecting out to the year 2100, the current trend would cause sea level to increase not by twenty feet, 9 feet, or even 2 feet (59 centimeters), but by 12.6 centimeters (5").*

The climate change community has been arguing for years that the oceans have been warming by as much as 2°F (1°C) (a gross exaggeration as shown in Chapter 6). Ironically, if this is true then even less ice has been melting than they claim. Water expands slightly when heated, as quantified by its *thermal expansion coefficient* of $7 \times 10^{-5}/°C$[33]. Even if one assumes that only the top 300 meters of the ocean is in thermal equilibrium with the atmosphere, and that both air and water temperatures have increased by 2°F over the past 100 years as claimed, the ocean should have expanded by around 4 cm, or 0.4 mm/year. After removing the contribution made by the thermal expansion of the ocean (4 cm per hundred years), the remaining sea level rise that can be attributed to the melting of ice is only 11 centimeters (4 inches) per hundred years.

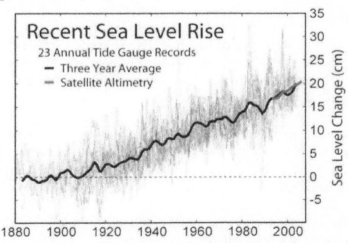

Fig. 5.8 Average sea level changes reported from 1880 through 2002 in centimeters (cm) compiled by the IPCC.[32]

If the rate of sea level rise since the end of the Little Ice Age continues unabated, Al Gore's predicted increase in sea levels by 20 feet could indeed come to pass - 4,000 years from now. The USGS prediction of a ten-meter rise in sea levels (Fig. 5.7) will occur in 6,700 years if another Ice Age doesn't happen first.

However, the next major Ice Age is anticipated in 5,000 years. As water levels were 18 feet higher than today just prior to the last major ice age, the current rate of rise is completely consistent with sea level changes that have accompanied previous (and completely natural) ice age cycles (see Chapter 3).

The Reality of Melting Ice

Finally, estimates regarding the melting of all major ice sources can now be reexamined based on the extent of sea level changes *that have actually been observed*. These data clearly show that the global warming community cannot have it all.

*The Earth is not experiencing a warming period that is sufficient to cause the oceans to expand **and** cause the ice fields in Antarctica to melt **and** induce the Greenland Ice Cap to melt **and** eliminate all temperate glaciers, ice shelves, and pack ice from the Earth. Actual sea level rises simply do not back up these claims.*

The actual total rate of sea level rise is at most 15 cm/100 years. If the warming of the oceans is to be believed, the total melt water generated by all global ice sources combined could contribute as little as 11 cm/100 years. Yet as shown in Table 5.1:

The melt water values used by climate change scientists and the media to describe the catastrophic disappearance of each individual ice source add up to 54 cm/100 years, or up to five times the sea level rise that has actually been observed.

Something is clearly amiss. To be generous, assume that there has been no warming of the oceans (i.e. no global warming?) and that the entire observed sea level rise is due to melting ice (but why

90

would it be melting if it isn't warming?). Starting with Antarctica, the upper estimate for the glacial melt along the west coast of the Antarctic Peninsula of 38 cm/100 years is clearly ridiculous. The maximum claim for this single source is almost three times higher than the actual melt water produced across the entire planet. Climate change scientists at institutions such as NASA have claimed that the melt water produced from Greenland and glaciers is up to 16 cm/100 years, which exceeds the total sea level rise that even the IPCC reports. Does this mean that there is absolutely no melting in Antarctica? A recent study in the journal *Nature* reported that between 2003 to 2010, Antarctica and Greenland have exhibited a combined melt rate of 10.6 cm/100 years, while all other ice sources contribute 4.1 cm/100 years for a grand total of 14.8 cm/100 years. This estimate is at least within range of the rise in sea levels that is actually observed.

In closing, Climate Change alarms regarding the melting of planetary ice due to the burning of fossil fuels are vastly overblown. At a minimum, actual increases in sea levels do not support the alarming climate change narrative that sea levels should mirror the exponential increases in fossil fuel emissions (Fig. 1.1). However, many scientists now believe that Fig. 5.8 represents an upper estimate for modern sea level changes, and that a more accurate value for the rate of rise is one centimeter per 100 years. In fact, recent evidence suggests that sea levels may not be increasing at all, but are currently falling.[34,35] A discussion of the ploys that have been used to inflate, explain away, manipulate, modify, and even fabricate sea level data to bring it in line with the climate change agenda are deferred to Chapter 9.

Humanity can relax. Coastal cities, civilizations, and planetary wildlife are not going to wiped out as a result of sea level rises amounting to only 6 inches per century. Fossil fuel combustion has had no impact on even this minor sea level rise. You do not need to be a scientist to know that this is true. Go to the beach. Go to any major coastal city. Have you noticed any changes in sea levels? Other than in isolated low-level locales such as Holland, do you see massive earthworks being constructed to protect our cities and infrastructures from the ice-melt induced floods to come? Humans and wildlife are adaptable, and have accommodated themselves to the 26 Ice Age cycles that have occurred during the past 2.5 million years since polar bears evolved. A thousand years ago, the Vikings colonized Greenland when the ice ebbed. Six hundred years ago, the Vikings abandoned Greenland when the ice advanced. The Vikings did not become extinct. Polar bears have survived quite well through all 26 Ice Age cycles. They did not become extinct either (see Chapter 7).) Each and every one of these cycles has produced dramatic changes in both ice and sea levels that dwarf anything that we are seeing today. These natural climactic cycles will continue into the future, eclipsing any impact on climate that humans might have.

Summary: Most of the vast quantities of ice on Earth are tied up in the massive ice caps in Antarctica (87%) and Greenland (13%), with the remainder residing in temperate zone glaciers, shelf ice, and pack ice. Although pack ice covers 6% of the ocean, it accounts for only 0.14% of all ice. The Antarctic Ice Cap is not melting. The Greenland Ice Cap is not melting. The pack ice is not melting. Almost the entire volume of melting ice resides in temperate and coastal polar glaciers. The total percentage of Earth's ice that has melted during the past 100 years is 0.16% (1/625). Since the end of the Little Ice Age in 1850, sea levels have been rising at a constant rate of at most 6 inches (15 centimeters) or less per hundred years. This rate has not been affected by the burning of fossil fuels.

Chapter 6: Weather and Climate

The Myth: The warming of the Earth's climate due to fossil fuel emissions is creating new and catastrophic weather patterns, including record heat and an increase in the number and intensity of hurricanes, tornados, droughts, floods, and even snowstorms.

Predictions of the Global Warming Movement

Americans receive a daily barrage from the media and climate experts reporting that each and every day, week, month, or year is the hottest on record due to global warming. The same climate experts warn that record heat is just the 'tip of the iceberg.' Americans are also told that global warming is the root cause behind any and all weather that is extreme, destructive, unusual, or uncomfortable. For example, *Boston Globe* columnist Ross Gelbspan wrote[1] that in addition to Hurricane Katrina, global warming was responsible for a blizzard in Los Angeles, high winds in Scandinavia, wildfires in Spain, and a drought in Missouri. Incessant messages include: 1) Americans should be afraid to go outside due to the new threatening weather conditions, and 2) Americans should feel guilty for burning the fossil fuels that are causing all of this climactic mayhem.

Is the Earth currently experiencing weather that is hotter, colder, wetter, or drier than ever before? Are violent storms including hurricanes and tornados gaining in both numbers and intensities to the point where the cost in damage and human suffering is too much to bear? How hot is the Earth after all of these years of record high temperatures? To evaluate these claims, one needs to examine actual weather records during the past hundred years or so to determine if rising atmospheric carbon dioxide levels are indeed leading to more extreme climactic conditions.

Temperature Records: Hot *and Cold*

Climate change advocates can always find somewhere on Earth where temperatures are hotter than ever. The focus is always on isolated temperatures that have reached all-time highs while ignoring any reports of all-time record lows. They would like you to believe that due to fossil fuel emissions, summers are now longer and hotter, while winters are shorter and milder. Do factual temperature records back up these claims?

To help evaluate media reports regarding isolated temperature records, it is useful to list what the Earth's temperature extremes have actually been in modern times.[2] Did Earth see its hottest temperature ever this year? The answer is no. The world record temperature ever reported was 136°F in Libya on September 13, 1922. The record high temperature for the United States was 134°F (Death Valley, CA, July 10, 1913). Fossil fuel emissions in both 1913 and 1922 were negligible compared to what they are today. The coldest temperature ever reported was -129°F in Vostok, Antarctica on July 21, 1983 when CO_2 emissions were five times higher than in 1913 when the U.S. heat record was set. The coldest temperature for a town in the lower 48 states (excluding Alaska) of -64°F was recorded as recently as February 2, 1996 in Embarrass, Minnesota. Did the media and climate scientists warn that this record low temperature indicated that we are headed for another Ice Age? The maximum reported difference between high and low temperatures at a single location is 188°F (from -90°F to 98°F) in Verkhoyansk, Siberia. In fact, there are over 22 cities in the United States that have set both record high and record low temperatures for a given day of the year within a single day. For example, on July 2, 1989, the temperature in Alamosa, Colorado varied between a low of 35°F and a high of 91°F for a temperature swing of 56°F. These examples all illustrate that cherry-picking record high temperatures at isolated locations tells you absolutely nothing about the Earth's climate.

Instead of looking at isolated temperatures, one can examine record temperatures set across an entire country or continent. Figure 6.1 represents a compilation of record high and low temperatures reported for every state in the United States as a function of the year in which the record was achieved.[3] Based on this compilation, 2016 is not even close to being the hottest year that America has experienced since 1900. The strongest heat wave ever recorded occurred in July of 1936, generating record high temperatures in half of America's fifty states. In 1935, fossil fuel emissions were 25 times lower than they are today. America's coldest year occurred in 1899, during which temperatures dropped below 0°F in all fifty states.

Record High and Low Temperatures by Decades and States

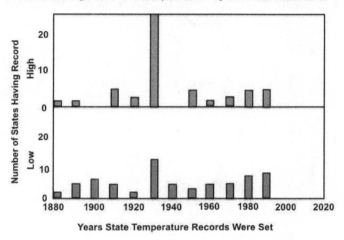

Fig. 6.1 Reports of record high (top) and low (bottom) temperatures by decade for states in the United States (from data in Ref. 3).

Interestingly, the most severe historic cold wave during the past 100 years took place in February of 1936, which was the same year when the strongest heat wave took place. In terms of general behavior, the global warming prediction is that as time progresses, and fossil fuel emissions increase, the number of record high temperatures should also increase, while the number of record low temperatures should decrease. No such trends are observed. Instead, *trends in record temperature data reported across all of North America dispel rather than support the global warming hypothesis.*

Rather than examining record high temperatures, one can examine the extent to which extended heat waves have taken place in modern times. For example, a compilation of all days since 1915 when average temperatures exceeded 90°F (Fig. 6.2[4]) show that *the number of hot days that Americans experience is decreasing rather than increasing.*

In this data, the heat wave experienced during the 1930s is clearly apparent, followed by another wave of high heat in the 1950s. No major heat waves have occurred since. However, there has been a major cold wave. The brutal winter of 2013 was the coldest and snowiest winter experienced across the entire United States in over thirty years. Once again, *documented heat wave data do not support the claim that the United States is hotter than ever, nor that rising carbon dioxide levels are causing global warming.*

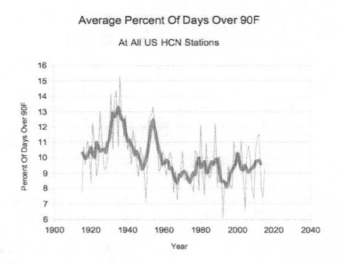

Fig. 6.2 The percentage of days for which temperatures exceeding 90°F were reported for all official stations in the United States.[4]

Of course, to really examine any shifts in climate, one needs to examine all temperature results from all reporting stations across the entire planet. Complete temperature records compiled prior to 1975 show that there have been four distinct shifts in global temperatures during the Industrial Age[5] (Fig. 3.8). (See Chapter 9 for recent attempts that have been made to try to falsify these climate records.) From 1880 to 1940, global temperatures increased by around 1°C (~2°F) to reach the highest values seen in the modern era. From 1940 to 1970, there was global cooling by around 0.7°C (1.3°F) that led environmentalists to warn of a coming Ice Age (see Chapter 1). From 1970 until 1998, there was the warming period by around 0.4°C (0.7°F) that helped spawn the Global Warming movement. Since 1998, little warming has occurred. All four phases are totally consistent with predictable variations in the amount of heat that the Earth receives from the Sun (see Chapter 3).

Accurate measurements of global temperatures have also been taken within the lower atmosphere where most weather and climate takes place. Continuous measurements within the mid-troposphere have been taken since 1979 using both balloons and satellite observations. Satellite data provide complete coverage of oceans as well as continents, providing a truly global perspective on the atmospheric temperature distributions. (Until recently, these satellite data had not yet been tampered with (see Chapter 9).) Satellite data provide confirmation to the two most recent of Earth's climate phases,[6] showing an increase by around 0.4°C (0.7°F) between 1978 and 1998 followed by the current era of stable temperatures (Fig. 6.3).

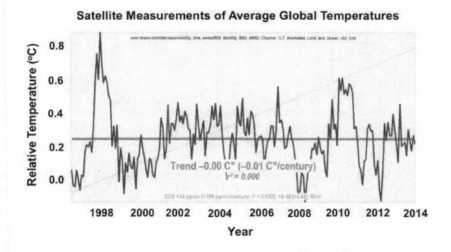

Fig. 6.3 Average global mean temperature changes[6] in °C obtained from satellite measurements between 1996 and 2013. The horizontal line indicates the average over all years in the data set. No temperature increase has occurred over the most recent time span.

The results in Fig. 6.3 clearly show that *the atmosphere has exhibited no warming for the past 19 years, in spite of the fact that carbon dioxide emissions have increased by a factor of four during that same time span.*

What about oceanic temperatures? An international array of over 4000 Argo floats have been programmed to collect temperature and salinity data of the ocean at any desired depth.[7] The average temperature change over the entire globe from the surface down to a depth of 700 meters is shown in Fig. 6.4. These near-surface measurements cover most seawater above the thermocline (see Chapter 4) where climate-induced temperature changes occur.

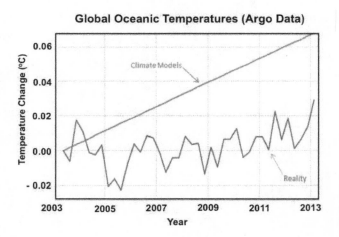

Fig. 6.4 Global oceanic temperature changes (bottom curve) covering the depth range from 0-700 meters as measured by Argo floats. Top line indicates climate model predictions. Data source: Argo 0-700m, NODC (National Oceanic Data Center), USA.[8]

Since 2003, the average slope of the curve represents a gradual increase of around 0.3°C per hundred years. This increase is almost identical to the average slope through land-based temperature records taken from the present back to 1880. The 'good news' for the environmental movement is that such a low temperature rise has been insufficient to induce the ocean to release enough CO_2 back into the atmosphere to account for modern increases in atmospheric concentrations (see Chapter 4). The 'bad news' is that the observed rate of 'climate change' is as much as ten times slower than claimed by climate change advocates (see Chapter 9).

Violent Storms: Hurricanes and Tornados

The year of 2005 was the worst hurricane season that the United States has suffered in recent memory.[9] During that year, 28 tropical storms were spawned in the Caribbean Sea and the Atlantic Ocean. Four of these storms were awesome Category 5 hurricanes that exhibited wind speeds over the ocean exceeding 155 miles per hour (mph). Three of these hurricanes made landfall. The most notorious of these storms was Hurricane Katrina. Katrina was so large that it almost filled the Caribbean Sea, packing maximum wind speeds of 175 mph and 125 mph (Category 3) on landfall in southeastern Louisiana near the city of New Orleans. The storm killed 1,300 people. It was the costliest storm in U.S. history ($108 billion), with most of the damage being associated with flooding due to storm surge and levee failures.

The hurricane season of 2005 came at an auspicious time for Global Warming, and has since become the weather 'poster child' of the movement. For example, Ross Gelbspan of the Boston Globe declared:[10] "Katrina's real name is Global Warming." The Katrina disaster arrived just in time for inclusion into Al Gore's famous movie *An Inconvenient Truth*. The recent hurricane season of 2017 has now supplanted 2005 in the public eye. Meteorological 'experts' including Leonardo DiCaprio, Stevie Wonder, and Pope Francis all claim that the 2017 hurricane season was unprecedented, vindicating predictions made back in 2005.[11] In a speech at the World Economic Forum in September, Al Gore led the charge with statements including:[11] Rainfall from Hurricane Harvey represented a "once in 25,000 year event" (see Precipitation Records below). "And why? Because today like all days we will put another 110 million tons of man-made heat-trapping pollution into the atmosphere, using the sky as an open sewer (see Chapter 7 for the actual toxicity of carbon dioxide)."

The narrative connecting Katrina to global warming involves the fact that heat provides the energy that drives both hurricanes and tornados. These violent storms form when columns of warm air rise and form swirling vortices in response to Coriolis forces associated with the Earth's rotation. Climate Change advocates argue that global warming should create columns of warm air that occur more frequently, and that are hotter, rise faster, and contain more energy. Therefore, global warming is predicted to increase both the number and intensity of violent storms. The standard narrative goes on to threaten that the 2005 and 2017 hurricane seasons were a mild taste of things to come. Climate change scientists and the media predict that so long as the burning of fossil fuels continues, the United States can expect to see more frequent and violent hurricanes, as well as more violent and frequent swarms of tornados across the mid-West.

It has now been thirteen years since the 2005 predictions were made regarding the dire fate the United States will face due to human-induced hurricanes and tornados. Are violent storms spiraling out of control? Do the severity of modern storms eclipse those that took place prior to the significant burning of fossil fuels? Starting with individual hurricanes, Hurricane Katrina was an impressive storm, but it only holds the record for being the most expensive, with much of that expense being caused by human mismanagement of the Louisiana levee system. The deadliest U.S. hurricane on record was the Galveston hurricane in 1900, which is thought to have killed over 8,000 people. The Great Hurricane in October of 1780 killed 22,000 throughout the Caribbean. The most intense hurricane in U.S. history was the Labor Day Hurricane of 1935 that slammed into the Florida Keys with wind gusts exceeding 200 mph on landfall. (This hurricane was highlighted in the movie *Key Largo* starring Humphrey Bogart and Lauren Bacall.) Neither storm could have been caused by fossil fuels, as emissions were still negligible.

Notice that individual hurricane record holders appear to be randomly distributed through time. To see if there is a pattern, one needs to examine the total number of hurricanes as a function of intensity and time.[12] Figure 6.5 depicts the number of severe hurricanes having wind speeds in excess of 110 mph that have formed over the Atlantic Ocean and the Caribbean Sea over the past 20 years.[12] On average, around two Category 3 storms and two to three Category 4 storms form per year. Over the past 20 years, only ten Category 5 storms have been produced, for an average of one every other year. In fact, 2017 was close to representing an average year in terms of overall hurricane activity, spawning two Category 3 and three Category 4 storms. The reasons 2017 seemed so active include: 1) incessant media coverage, 2) there was essentially no hurricane activity for the previous five years, and 3) two of the Category 4 storms (Harvey and Irma) made landfall on the continental United States. Conversely, only one Category 5 storm has been seen during the past ten years. Even including 2017, only 12 out of the 61 hurricanes in all three categories (20%) have occurred during the past six years, which is substantially fewer than the 20-year average. This highlights how truly unusual the 2005 season was in generating four of these ten most massive storms.

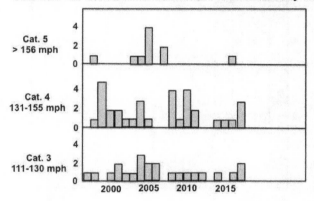

Fig. 6.5 The number of hurricanes observed over the Atlantic Ocean and the Caribbean Sea between 1995 and 2015 as a function of hurricane intensity (from data in Ref. 12).

To summarize:

In spite of 2017, none of the predictions made by the global warming movement regarding the number, intensity, or impact of hurricanes on the United States have come to pass.

In discussing tropical storms, the typhoons and cyclones that form over the Pacific and Indian Oceans cannot be ignored. In fact, typhoons are much more significant than hurricanes in terms of their numbers, intensities, and impact on human life. The largest documented tropical storm was Super-typhoon Tip, which formed over the Pacific in October of 1979. The deadliest tropical storm in terms of loss of life was the Bangladesh storm of 1970, which is estimated to have killed from 300,000 to 500,000 people. Again, this storm provided momentum to the global warming movement. Was this storm a portent of things to come? Once again, the answer is no. Figure 6.6 shows the years during which the thirty deadliest typhoons occurred[12] (organized in increments of twenty years).

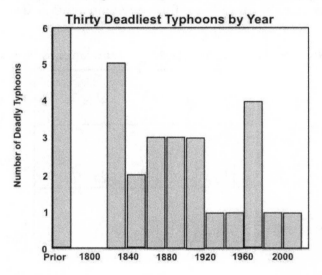

Fig. 6.6 A bar graph showing when the thirty deadliest typhoons in known history occurred, with the most recent (since 1820) being compiled in 20-year intervals. (Data from Ref. 12.)

While the decades of the 1960s and 1970s (near the end of the most recent period of *global cooling*) were particularly active, the general trend all the way back to 1820 has been a decline in the number of deadly typhoons. Since 1980 there have only been two (6.7% of the total). In other words:

The number and intensity of deadly typhoons has been decreasing while carbon dioxide emissions have increased by a factor of ten.

The final class of violent storms to consider is tornados. In terms of their size, tornados are much smaller than hurricanes, rarely exceeding one mile in diameter. However, in terms of wind speeds, tornados represent the most extreme storms on Earth. Tornados are classified according to their wind speeds[13] on the Fujita (or F) Tornado Scale. The three most intense rankings are F-3 (158-206 mph), F-4 (207-260 mph), and incredibly frightening F-5 (261-318 mph) storms. In terms of devastation, F-3 tornados are capable of ripping the roofs off of houses, overturning trains, and uprooting trees, F-4 tornados can level well-constructed houses and throw cars, and F-5 tornados can lift complete houses off of their foundations, hurl cars distances of over a hundred yards, and even destroy steel-reinforced concrete structures.

Are tornados currently more numerous and intense than ever due to global warming? Regarding single tornados, climate change advocates can point to the deadliest tornado in modern history,[14] which occurred in Bangladesh on April 26, 1989. This storm killed 1,300 people and left 30,000 people homeless. However, as a counter example, the deadliest tornado in U.S. history occurred on March 18, 1925 before the era of significant carbon dioxide emissions. Advocates can also point to April 3-4, 1974 that set the U.S. record for the most tornados (148) spawned within 24 hours. However, the best measure of tornado activity involves counting all of the tornados that form within a given year. In terms of both numbers and intensities, the United States is the dominant country in the world when it comes to tornado generation. Since many more tornados form each year than hurricanes, it is also possible to obtain a more statistically significant sampling of violent storm activity from tornado data. Figure 6.7 depicts the number of truly violent tornados (F-3 or higher) that have been counted in the United States each year since 1954.[15] With the exception of a few anomalous years, such as 1974 and 2011, the tornado count shows a clear trend. The trend is downward.

Although carbon dioxide emissions are currently ten times what they were in 1954, there has been as much as a 25% drop in the numbers of truly violent tornados over the past fifty years.

In summary, there is not a single class of violent storms whose incidence or intensity has increased in response to dramatic increases in atmospheric carbon dioxide emissions (Fig. 1.1).

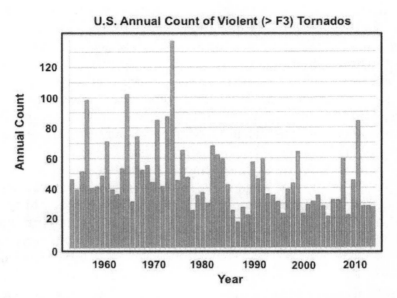

Fig. 6.7 The total number of strong to violent tornados observed in the United States from 1954 to 2014. Adapted from Ref. 15.

Precipitation Records: Floods and Droughts

The component of weather having the greatest impact on humanity is precipitation in the form of rain and snow. Too little precipitation leads to droughts and crop failures, while too much leads to flooding and the loss of property and life. In fact, more deaths are due to flooding than any other weather related cause. However, just like record high temperatures, record rainfall and flooding are still unpredictable random events. Anyone who examines rainfall records for both the United States (Fig. 6.8) and the world[16] can readily determine that there is no relationship whatsoever between record rain storms and carbon dioxide emissions.

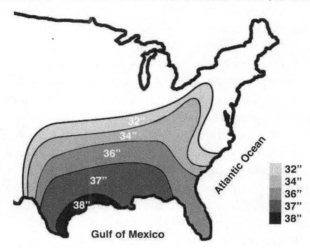

Daily Rainfall Records and Predicted Maximum Rainfall (Eastern U.S.)

Fig. 6.8 A map of the mid-western to eastern portions of the United States showing the maximum rainfall that hydrologists predict can occur within a single twenty-four hour period. Actual rainfall records reported within each zone are: Zone 1 (> 38") = 43", Zone 2 (37-38") = 32", Zone 3 (36-37") = 39", Zone 4 (34-36") = 24", and Zone 5 (32-34") = 35." Adapted from p. 117 in Ref. 16.

Al Gore's statement that the weekly record for rainfall from Hurricane Harvey constituted a "once-in-25,000 year event" caused by global warming has absolutely no basis in fact.

While it is true that Harvey established a new record for the continental U.S. at one reporting station of 48.2 inches for the week, this record was not set because Harvey was the fiercest storm ever. It was because a region of high pressure caused Harvey to stall out above Houston rather than continuing to move across the land as most hurricanes do. The two heaviest days of rain associated with Harvey delivered 8 inches on August 26 followed by 16 inches on August 27. However, while 24 inches in two days is impressive, eleven U.S. states have received more rain than that *within a single day*. For example, Alvin, Texas received 43 inches on September 25, 1979. The other daily records by state are randomly scattered between 1921 and today. Globally, Al Gore won't have to wait 25,000 years to see rainfall exceeding the weekly total for Hurricane Harvey. In fact, the rainfall record *for a single day* of 73.6 inches was set in 1952 on Reunion Island in the

Indian Ocean. Since 1876, record daily rainstorms of 40 inches or more have occurred every seven years or so. The weekly record (also at Reunion Island) of 183 inches set in 1980 is almost four times greater than the Harvey total.

Anyone who watches the nightly weather forecast on television knows that weather is an extremely complex phenomenon. Even trained meteorologists, who have access to the latest satellite photographs, Doppler radar, and input from other weather stations, have a hard time predicting what a given storm will do. For example, on March 12, 2017, a major winter storm called a nor'easter slammed into the entire northeast coast of the United States from Virginia through New England.[17] The storm was predicted to dump 24 inches of snow in New York City, a foot of snow in Boston and Philadelphia, and 8 inches in Baltimore, triggering the cancellation of 6,100 airplane flights and the closure of schools throughout the East Coast. Unfortunately, most of these airline cancellations and school closures were unnecessary. The actual storm was much less severe than predicted, producing only 4-6 inches of snow across New York City and Philadelphia. As a result millions of dollars and untold hours of productivity were squandered to combat a storm that never really materialized.

Although the behavior of individual storms is hard to predict, meteorologists claim to be making progress when it comes to forecasting extended weather patterns over extended geographical regions. The primary factor evoked to predict both precipitation and temperature patterns across much of the world is a phenomenon called the El Nino-Southern Oscillation (ENSO).[18] This phenomenon is driven by temperature anomalies that occur within a vast stretch of the eastern Pacific that straddles the Equator ($\pm 5°$ in latitude) 1900 miles south and southeast of Hawaii. If oceanic surface temperatures within this zone increase by 0.5°C (0.9°F) or more, a weather pattern called El Nino is predicted to occur. Conversely, a drop in oceanic temperatures within the zone by 0.5°C triggers a pattern called La Nina. Typically, oceanic temperatures oscillate between hot and cool phases every 9 months to a year. Particularly strong El Nino or La Nina events are typically observed every 15 years or so.

Anomalies in the temperature of the eastern Pacific Ocean impact the weather because they drive atmospheric pressure differences across the entire Pacific Ocean. Much as regions of high and low atmospheric pressure drive storm tracks across the United States, differences in atmospheric pressure across the Pacific determine air circulation patterns over much of the Earth. Pressure differences measured between the island of Tahiti to the east and Darwin, Australia to the west are used to calculate a quantity called the Southern Oscillation Index (SOI). SOI values less than -10 are thought to signal strong El Nino air circulation patterns, while values greater than +10 signal La Nina circulation.

Weather patterns for El Nino and La Nina circulation systems are predicted to be dramatically different in different regions of both the United States and the rest of the world. For example, much of Southeast Asia is predicted to be hot and suffer through droughts during El Ninos, while experiencing flooding and cooler temperatures during La Ninas. Figure 6.9 shows the pattern predicted for extended winter weather across the United States during El Ninos.[19] A strong El Nino is expected to produce wetter, cooler weather across the southern United States from California all the way to Florida. Conversely, warmer, dryer weather is expected all the way from the Pacific Northwest to the Ohio Valley near the east coast. A strong La Nina pattern is expected to exhibit the opposite trends. The climate change narrative is based on the premise that global warming should produce warmer conditions in the Pacific, resulting in more numerous and intense El Ninos and a suppression of La Nina events. Advocates claim that the past several decades confirm that El Ninos are beginning to dominate. This means that the United States should experience longer and more severe droughts in the mid-west with dire consequences for the nation's food supplies. To back this claim, the global warming community points to the severe drought that struck Missouri in 2012, which was the worst U.S. drought in a generation.

Are El Nino's spiraling out of control due to global warming? Are we seeing the most severe weather conditions that the U.S. has ever seen? Are Americans threatened with such severe droughts that the starvation of our citizens is immanent? An examination of all weather conditions and trends that have been reported across the United States can determine if this is true.

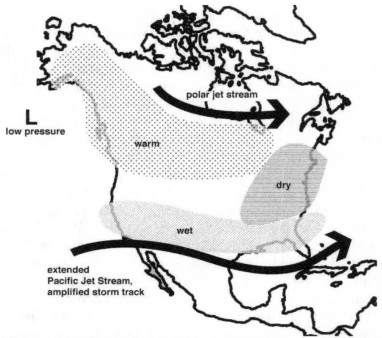

Fig. 6.9 A map of the United States showing regions of the country expected to be warm, wet, or dry as a result of a wintertime El Nino pattern. Adapted from Ref. 19.

First, examine what El Nino and La Nina have actually been doing. Figure 6.10 depicts the magnitude of the Southern Oscillation Index going all the way back to 1880.[20] Here, negative deflections depict El Nino events, while positive deflections represent La Nina's. No long-term trends are apparent. If you look hard, you can see that El Ninos outnumbered La Ninas between 1980 and 2000 in line with global warming claims. However, you are not supposed to notice that La Ninas were predominant between 1940 and 1970. You are absolutely not supposed to notice that since about 2005, La Ninas are more common during a period of time when fossil fuel emissions have more than tripled.

The bottom line is: *Observed changes in the Southern Oscillation Index are consistent with changes in solar activity rather than on increasing fossil fuel emissions. The predicted pattern of runaway El Nino behavior that tracks carbon dioxide emissions has not occurred.*

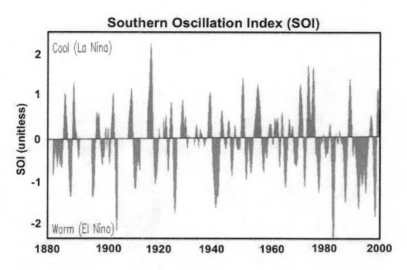

Fig. 6.10 A linear record of the number and intensity of El Nino (below zero) and La Nina (above zero) events from 1880 to the present based on the Southern Oscillation Index. Data are adapted and smoothed from information presented in Ref. 20.

Second, one can evaluate whether El Nino and La Nina behaviors actually control precipitation patterns across the United States. Figure 6.11 is a subset of the maps selected by NOAA (the National Oceanic and Atmospheric Administration) to show that the wintertime precipitation in the United States is dictated by El Ninos.[21] What the maps actually show is that short-term climate forecasting is still almost as uncertain as daily weather forecasting. The precipitation map that comes closest to matching El Nino predictions (Fig. 6.9) corresponds to the strong 1997 El Nino (upper left in Fig. 6.11), as precipitation was higher than normal in both California and the South, while falling below normal in parts of the Great Plains. However, the rest of the maps show how fickle the actual climate can be. California has experienced both flooding and droughts during both strong and moderate El Ninos, as has the South. There have been El Ninos during which California was wet while the South was dry, as well as El Ninos during which California was dry while the South was wet.

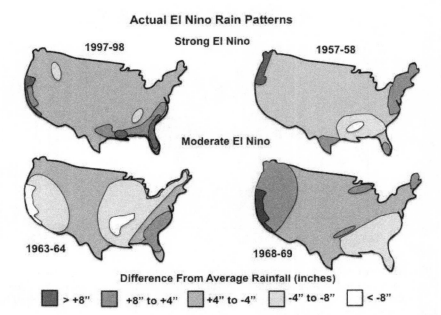

Actual El Nino Rain Patterns

Strong El Nino

1997-98   1957-58

Moderate El Nino

1963-64   1968-69

Difference From Average Rainfall (inches)

> +8"   +8" to +4"   +4" to -4"   -4" to -8"   < -8"

Fig. 6.11 Maps of the United States showing regions of the country that experienced more or less precipitation than normal (see lower bar) during strong or moderate El Nino events. Adapted from maps presented in Ref. 21.

In short: *The United States exhibits no regular or predictable precipitation patterns that clearly depend on El Nino behavior.*

As ENSO predictions don't appear to be a strong indicator of precipitation, one needs to examine the actual periods of both wet and dry weather conditions that the United States has experienced during the past 100 years. One means of doing this to examine those years reporting the most and least rainfall per state within a calendar year (Fig. 6.12).[22] Did the recent drought of 2012 set any records? The answer is no. The worst drought (and perhaps the worst natural disaster) in United States history was the great Dust Bowl event that occurred between 1933 and 1938 (corresponding to the heat wave shown in Fig. 6.1). This drought encompassed over 50 million square acres in a region stretching from Texas to Canada and from Colorado to Illinois. During the drought, the topsoil was so dry that crops did not grow, and massive windstorms blew much of it away.

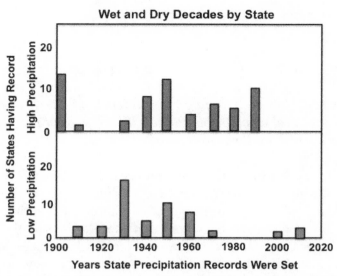

Fig. 6.12 A bar graph showing years (by decade) in which wet (top) and dry (bottom) precipitation records were set for states within the United States (from data compiled in Ref. 22).

The nation's second worst drought occurred in 1949-1951, during which the Great Plains and the Southwest experienced a 40% drop in annual rainfall. Interestingly, the greatest number of states exhibiting their highest precipitation levels occurred shortly afterwards (in the mid-to-late-1950s). Over the entire period from 1900 to today, there doesn't appear to be any regular or predictable trends regarding whether states are getting wetter or drier. The wettest decades are clustered between 1940 and 1990. However, the number and intensity of droughts appears to have undergone a decline since around 1960. In fact: *Droughts have decreased while carbon dioxide emissions have increased by a factor of ten.*

Summary: The Earth's climate and weather patterns have been remarkably stable over the past 100 years. It is not significantly hotter, colder, wetter, or drier than it was before the era of fossil fuel combustion. If anything, the incidence of violent storms including both hurricanes and tornados is down rather than up. Cherry-picking isolated extreme weather events in an attempt to claim otherwise represents a dishonest and irresponsible strategy aimed at frightening people into believing the false premise that fossil fuel emissions are destroying the planet.

Chapter 7: The Impact of Global Warming on Life

The Myth: "Global Warming . . . is causing the loss of living species at a level comparable to the extinction event that wiped out the dinosaurs 65 million years ago." – Al Gore[1]

Is all life on Earth facing obliteration due to Global Warming? Will a warming of the planet help or hurt most of the Earth's ecosystems? What life forms are actually threatened with immanent extinction? It has been estimated that there are currently over 10 million species of living organisms on Earth.[2] There are more species in existence now that there have ever been. New species are constantly replacing old species. Extinctions have always been an integral part of the evolution of life on Earth. In terms of the global warming movement, central questions include: 1) whether the current rate of extinctions is highly unusual, 2) whether humans are causing these extinctions, and 3) whether any of these modern extinctions can be directly tied to global warming. Answers to these questions are provided below through a consideration of: 1) factors that contribute to extinctions, 2) the impact that carbon dioxide and temperatures have on life, 3) over 600 million years of the geologic record regarding the impact that both CO_2 and temperature changes have had on life and on extinctions, 4) information regarding what species are currently in danger of extinction due to either human activities or global warming, and 5) historical and archeological records showing the impact that warmer and colder climates have had on human life.

Causes of Extinctions

A range of interrelated phenomena can contribute to extinctions including[3]:

Environmental Toxicity: Global changes in the composition of the air or water can transform the environment from a benign to a toxic condition. An example of such a transformation in the Earth's early history involves the conversion of life-supporting CO_2 into oxygen gas that was toxic to early single-celled photosynthetic organisms[4].

Drastic Temperature Changes: Many species are adapted for life within a specific climate. Polar species cannot survive in the tropics, nor can tropical species survive exposure to extreme cold. In temperate climates, species are adapted to specific annual temperature cycles. Species are most vulnerable when the climate changes over a period of time that is too short to allow them to adapt to new conditions. The most famous example of a mass extinction caused by climate change involves the disappearance of the dinosaurs. Many scientists believe that this extinction was due to a rapid drop in planetary temperatures resulting from an asteroid impact[5] (see the Geologic Record below).

Habitat Destruction: Specific species are adapted to survive in specific ecosystems. When an ecosystem is destroyed, all species in that ecosystem are at risk. For example, some environmentalists have predicted that up to 21% of forest species in Southeast Asia will disappear by 2100 due to massive ongoing deforestation.[6]

Competition: The introduction of invasive species that either eat or compete with native species for food or habitat can cause extinctions. For example, the inadvertent introduction of the brown tree snake to the island of Guam during World War II is thought to be responsible for the loss of 12 of the 18 native bird species on the island.[3]

Invasive Diseases: Exposure to foreign diseases can be devastating, particularly for small, isolated populations. Unless a sufficient population exists to develop immunity to the new disease, extinction can occur. A human example of such a pandemic is the Black Death[7] (bubonic plague) that migrated from China to Europe in 1347 and killed up to 200 million people, or almost half of the world's population at the time.

Reproductive Failure: The above conditions have the greatest impact on creatures having reproductive rates that are too low to keep pace with increased population losses. Examples include the extinctions of the woolly mammoth and saber tooth cat in North America.

Examining the above list, there is no question that human activities have contributed to extinctions. As human populations have expanded, humans have increasingly come into competition and conflict with a wide range of animals and plants. Humans are omnivores that can eat almost anything. Some organisms are killed for food, while others are killed that compete for human food sources. Animal habitats are routinely destroyed. Humans inadvertently introduce new species and diseases into previously isolated environments, creating non-human threats to native species. However, *none of these unfortunate occurrences have anything to do with global warming.* The key question regarding the global warming movement is the extent to which the introduction of carbon dioxide and other greenhouse gases either poisons plants and animals or causes temperature changes that are sufficient to cause extinctions.

The Impact of Carbon Dioxide on Life

The most ridiculous claims made by the Climate Change movement fall under the heading that carbon dioxide is destroying all life on Earth. The exact opposite is true:

Life as we know it cannot exist without carbon dioxide.

Photosynthesis forms the basis for life on Earth. Carbon dioxide forms the basis for photosynthesis. Photosynthesis is the mechanism that plants, algae, and cyanobacteria use to convert light into chemical energy.[8] Photosynthesis provides the basis for all of the organic compounds and most of the energy used by most of Earth's food chains via reactions such as:

$$6 \, CO_2 + 6 \, H_2O + light \rightarrow C_6H_{12}O_6 \text{ (sugar)} + 6 \, O_2 \quad (7.1)$$

Photosynthesis produces around 1,100 terawatts (trillion watts) of biological energy every year,[9] which is forty times the total annual energy use of the United States (see Chapter 8). Photosynthesis creates over 100 billion metric tons of biomass per year. The origin of all carbon in this biomass is carbon dioxide. The Earth's entire food chain is based on carbon dioxide. Yet Climate Change advocates claim that CO_2 is an evil toxic gas that needs to be removed from the environment.

Go figure.

Carbon dioxide is called a 'greenhouse gas' for a reason. As discussed above, most plants cannot grow or survive without CO_2. Carbon dioxide is the primary nutrient consumed during photosynthesis and is the source for all organic carbon. Horticulturists deliberately inject CO_2 into greenhouse air to stimulate plant growth. Some environmentalists have even advocated placing greenhouses near fossil fuel plants to siphon off and consume the CO_2 emissions for use in growing plants. Beneficial effects of CO_2 include a shortening of growing periods, increased plant yields, and increased crop quality.[10] The effect of CO_2 on plant growth is presented in Fig. 7.1.

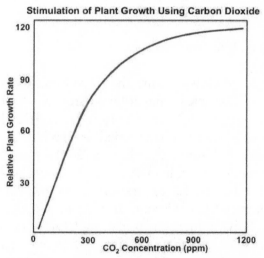

Fig. 7.1 Relative rates of plant growth as a function of atmospheric carbon dioxide concentrations (in ppm). From data in Ref. 10.

At a concentration of zero CO_2, plants do not grow at all. Growth rates are directly proportional to CO_2 concentrations from zero up to around 300 ppmv. For most plants, the optimum concentration for maximum growth is 1200-1300 ppmv, or three to four times greater than existing atmospheric concentrations (see the Geologic record below). This concentration would not even be reached if all fossil fuels on Earth were incinerated (see Chapter 4).

The beneficial effects of CO_2 start to drop above 2000 ppmv where CO_2 begins to interfere with plant respiration, resulting in toxic side effects. In respiration,[8] plants generate chemical energy by reacting a fraction of the sugars that they produce via photosynthesis with oxygen:

$$C_6H_{12}O_6 + 6\ O_2 \rightarrow 6\ CO_2 + 6\ H_2O + energy \qquad (7.2)$$

Respiration reactions in plants are similar to the respiration reactions used by animals to generate energy, and are the reverse of photosynthesis. The net amount of carbon incorporated into plants during growth represents the difference between the photosynthetic and respiration reaction rates. Note that the CO_2 concentrations needed to interfere with plant growth and health are 5 times greater than existing atmospheric concentrations.

The Toxicity of CO_2 to Humans

The behavior of zealots within the 'Environmental Protection' Agency (EPA) illustrates the total disregard that the global warming community has for obvious scientific facts. Under President Obama, the EPA succeeded in having carbon dioxide labeled as a 'toxic pollutant.'[11] This enabled them to regulate fossil fuel emissions under the Clean Air Act to try to justify shutting down all coal-fired electrical generation plants (see Chapter 8).

Just how toxic is CO_2? Carbon dioxide is a colorless, odorless, tasteless, and relatively inert gas. Labeling CO_2 as a toxic substance makes as much sense as labeling water as a poisonous chemical. It is true that both carbon dioxide and water can cause death due to suffocation (or drowning) by reducing the supply of oxygen to the lungs. However, carbon dioxide is just slightly more toxic than inert gases such as helium, argon, or nitrogen. Suffocation occurs if any of these gases dilute the oxygen in the air from its natural concentration of 21% to below 19.5%. The Center for Disease Control (CDC) has determined that the atmospheric concentration of CO_2 has to exceed 50,000 ppmv (5%) to be immediately dangerous to humans.[12] This concentration is 125 times greater than current atmospheric concentrations. A concentration of 100,000 ppmv (10%) leads to unconsciousness and death. In terms of worker safety, a concentration of 5000 ppmv is reported to produce symptoms such as dizziness and a feeling of intoxication after 30 minutes of exposure in a confined space. Fresh air immediately eliminates these symptoms. Note that carbon dioxide is not to be confused with its toxic cousin carbon *monoxide (CO)*, which becomes dangerous at concentrations exceeding only 1200 ppmv.[13]

The Impact of Temperature on Life

Of course the major argument that the global warming community uses to justify the banning of fossil fuels is that it is causing a runaway increase in global temperatures. It is claimed that these temperature increases are causing mass extinctions. Problems associated with these claims are: 1) the Earth's temperature is not increasing beyond natural temperature cycles that have been insufficient to cause extinctions in the past (see Chapters 3 and 9), and 2) carbon dioxide emissions do not control the Earth's temperature (see Chapters 2 and 4). In addition, climate change advocates in the 1970s got one thing right:

Global cooling represents a much greater threat to life on Earth than global warming.

Global Warming
Tropics

Global Cooling
Arctic

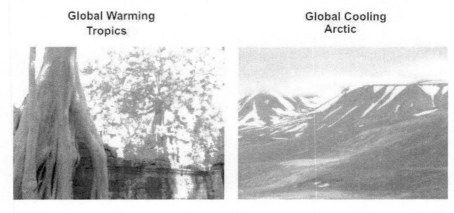

Fig. 7.2 Photographs of plant life in warm and cold climates. Left – Ruins at Angkor Watt in tropical Cambodia that are being reclaimed by the jungle. Right – Iceland in August. Although it lies just below the Arctic Circle, the climate of Iceland is too severe to support any native trees. (Author photos.)

You do not have to have an advanced degree in biology to reach this conclusion. You only need common sense. On land, the entire food chain is based on plant life. Do plants grow better in warm or cold climates? Is there more abundant and diverse life in the tropics or at the poles (Fig. 7.2)? Do crops grow better in the summer or in the winter?

Tropical forests cover less than 12% of all land, yet they contain over 50% (half) of all of the 10 million plant and animal species that inhabit the entire Earth.[14] In sharp contrast, while the Arctic encompasses 10% of the land area, few species can survive the harsh conditions of the frozen tundra.[15] There are only 600 species of hardy ground-hugging plants in the Arctic. Conditions are too cold to allow any reptiles or amphibians to survive. The Arctic contains only 100 species of birds, most of which are only present during the short summer nesting season. The list of mammals is so short (20) that half of them can be listed here: caribou, musk oxen, Arctic hares, lemmings, polar bears, brown bears, wolves, Arctic foxes, wolverines, and ermines.

Climate change advocates try to circumvent the obvious connection between warm climates and biodiversity by claiming that the secondary effects of global warming such as droughts or the melting of all planetary ice will extinguish existing life. Unfortunately for their argument, there has been no increase in the number or severity of droughts since the start of the Industrial Age (Chapter 6). Ice is not disappearing at an alarming rate (Chapter 5), nor has polar life been disappearing with it (see Polar Bears and Walrus below).

The Geologic Record

Proof that higher atmospheric carbon dioxide concentrations and/or temperatures are not endangering life on Earth is evident over at least 600 million years of the Earth's history.

The geologic record shows that temperatures and CO_2 levels have been higher than they are today throughout almost the entire history of life on Earth (Chapters 3 and 4).

The terrestrial food chain is based on plants. What conditions stimulate plant growth (see above), and what conditions existed on Earth when plant life was the most prolific? In every instance, plant life thrived when temperatures and carbon dioxide levels were both substantially higher than they are today (Fig. 7.3).[16]

Over 50% of all coal reserves on Earth represent decayed plant matter that was deposited during the aptly named Carboniferous Period[17] (360 – 290 million years ago or Mya). During much of this Period, temperatures were 15°F hotter than they are today. Carbon dioxide levels during the period started above 2000 ppmv and gradually dropped to modern levels of 400 ppmv in part due to the photosynthetic consumption of CO_2 by plants. In other words, atmospheric conditions were similar to those found in modern greenhouses. In fact, most plants evolved under greenhouse conditions. The thickest known coal beds were deposited during the early Eocene[17] (56-36 Mya) when temperatures were almost 20°F hotter than today. During the Eocene, tropical jungles were present even in polar regions. Carbon dioxide levels were 800 ppmv, or twice what they are today.

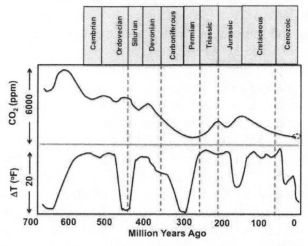

Fig. 7.3 Climactic conditions during the geologic eras of the Earth (top). The top graph shows carbon dioxide concentrations (in ppm), the bottom graph shows temperatures (in °F), while the dashed lines indicate major extinction events. From data in Ref. 16.

Similar trends are seen for animal life. How did the dinosaurs get so big? The answer is that they had enormous quantities of food available to support their bulk. Gigantic insects ruled the Earth during the tropical Carboniferous Period, such as six-foot long millipedes and dragonflies with a wingspan almost as wide. The dinosaurs appeared during the tropical Triassic and Jurassic Periods (250 – 150 Mya) when temperatures were 15°F hotter than today and CO_2 levels averaged around 1000 ppmv. The era of large mammals was launched during the tropical Eocene Period when horses, elephants, whales, rhinos, and extinct species such as giant sloths first made their appearance.

If tropical conditions allow life to thrive, what conditions lead to extinctions? More than twenty mass extinctions have been documented in the geologic record,[18] the most severe of which are shown as dashed lines in Fig. 7.3. The most devastating mass extinction occurred at the end of the Permian period around 250 million years ago (Mya). The Permian extinction is the only known Category 1 extinction event, which removed more than 50% of the existing families and up to 95% of all species on Earth. There have been four definite Category 2 extinctions during which 20-25% of all families and up to 50% of all species disappeared. These include events that occurred in the late Ordovician (438 Mya), the late Devonian (367 Mya), the late Triassic (208 Mya), and the famous extinction of the dinosaurs, which occurred at the Cretaceous-Tertiary boundary (65 Mya).

No climactic explanation is known for either the Devonian or Triassic extinction events. However, all three remaining Category 1 and 2 extinctions have been attributed to global cooling.[19] The Ordovician extinction (438 Mya), which is thought to have destroyed up to 85% of existing families, is apparent in the geologic record as an extended period of heavy glaciation and lower sea levels. The most catastrophic Permian extinction is associated with a short but intense period of glaciation (277 – 268 Mya). This temporary glaciation was apparently caused by sulfur dioxide emissions associated with the most intense period of volcanic activity experienced by terrestrial life, which led to the formation of the 600,000 square miles of lava known as the Siberian Traps[20] (see Chapter 3). The Siberian Trap eruption took place on a time scale that was too short (anywhere from 65,000 to one million years) to show up in Fig. 7.3. Similarly, the extinction of the dinosaurs coincided with the second largest known eruptive period that formed the Deccan Traps in India. It is thought that cooling due to this eruption set the stage for an extremely short (ten years or so) 'global winter' that was triggered by an asteroid collision 65 Mya.[21] The bottom line is that while scientists have yet to identify a major mass extinction associated with global warming, three of the largest extinction events in Earth's history have been attributed to global cooling.

Endangered and Threatened Species

What species are in danger of becoming extinct, and which of these species are threatened due to global warming? As noted above, Al Gore has implied that modern extinctions are due to global warming, claiming that the current extinction rate is equivalent to the Category 2 event that wiped out the dinosaurs. As a point of reference, the Earth is currently estimated to contain over 10 million species. To qualify as a Category 2 event, human-induced global warming will have to exterminate 5 million of these species.

Conservation organizations routinely rank species relative to their risk of extinction.[22] In descending order of the threat, species are: critically endangered, endangered, vulnerable, near threatened, or of least concern. Threat levels are determined by considering a wide range of factors, including existing and historical populations, whether populations are increasing or decreasing, and whether habitats are being destroyed. As examples, mountain gorillas, with only 400 individuals, and hawksbill turtles, with a population estimated at 25,000, are both on the critically endangered list. Snow leopards (population = 6,000) and sea lions (population 50,000) are both considered to be endangered.

The good news regarding conservation rankings is that many threatened species have seen significant population gains and a reduction in threat level due to enhanced public awareness and protection. For example, the worldwide population of blue whales[23] (10,000 – 25,000) has more than doubled since a whaling ban was instituted in 2002. The bad news regarding the rankings is that they have become highly subjective. Estimates for the number of threatened species range from a few thousand to over 100,000. For example, the U.S. Fish and Wildlife Service (May 2017) compiled a list of 1,845 plant and animal species that they consider to be either endangered or critically endangered throughout the world.[24] Unfortunately, their tabulation has to be incorrect at least in terms of foreign species. They list only one endangered foreign plant in spite of the fact that the most intense deforestation is taking place in Africa, Southeast Asia, and South America. Perhaps threats to foreign species are being downplayed in an attempt to make it appear that the United States is responsible for all extinctions when in fact America is a world leader in wildlife protection.

Most endangered species are large animals having relatively low reproductive rates. At this point in time, the World Wildlife Federation (WWF) has identified 19 major animal species that are critically endangered.[2] All of the critically endangered animals are threatened due to human predation and habitat destruction. For example, the orangutan population of 15,000 animals is dwindling as a result of deforestation. Not a single one of these species is endangered due to global warming. The WWF has identified 27 animal species that are endangered. Again, every species on the endangered list has suffered either due to hunting or habitat destruction. Not a single one is endangered due to global warming. In fact, Galapagos sea lions have seen their numbers grow from 20,000 up to 50,000 animals since 2002. In the next lower category of threatened animals, the WWF lists 20 animal species that are vulnerable. Once again, hunting and habitat destruction dominate the threats to animals in this group *with one possible exception*. Climate change advocates claim that polar habitats are being destroyed due to global warming, leading to the pending extinction of Arctic species such as the *polar bear*. Will polar bears really cease to exist in our lifetime due to global warming?

Polar Bears

The 'poster animal' of the climate change movement is the polar bear. The polar bear is an efficient carnivore at the top of the Arctic food chain. Climate change advocates always depict polar bears as cute and cuddly creatures stranded on melting ice floes rather than as a killer and eater of seals, in order to appeal to children and the emotions. However, the main reason that the climate change movement is constantly pointing at the polar bear is that it is the most visible species whose population might be at risk as a result of global warming.

The climate change community uses the following chain of logic to 'prove' that global warming is leading to the extinction of the polar bears. They claim that:

- Global warming is melting all of the pack ice on Earth.
- Pack ice represents the primary habitat of polar bears.
- As polar bears cannot survive without pack ice, global warming must be destroying polar bears.

Unfortunately for the global warming movement, actual facts regarding polar bears do not substantiate these claims.[25]

- Polar bear habitats will not be disappearing anytime soon. In spite of Al Gore's prediction that all pack ice would be completely gone by 2012, the extent of the pack ice has not changed significantly since comprehensive satellite imagery of the Arctic began back in 1981 (see Chapter 5). Two thirds of the pack ice melts each summer and reforms back to its starting point each winter without wiping out polar bear populations.

- Polar bears do not need pack ice to survive. Polar bears have been in existence since the current period of Ice Ages commenced 2.5 million years ago.[26] During that time, the Earth has experienced 24 major Ice Age cycles (see Chapters 3 and 5). During each major cycle, the Earth's temperature has been both substantially hotter and colder than it is today. At some point in each cycle, there has been much less pack ice than there is today. The polar bears have come through each of these cycles unscathed. They have not become extinct.

- Polar bears are not extremely endangered. They are not even considered endangered. The species is commonly listed as vulnerable, having a status comparable to that of the African elephant, the hippopotamus, and the marine iguana.

- Polar bear populations are increasing rather than declining. The current population is estimated to be 28,000 individuals.[25] This population is almost three times higher than the estimated 10,000 animals that were present in 1960. The 1960 population was low due to overhunting. After application of hunting restrictions, the population increased to a new healthy level above 25,000 individuals between 1960 and 1990.

Of course the media refuses to report on the fact that polar bear numbers are robust. By definition, polar bear numbers must be decreasing at an alarming rate. Unfortunately, the only means the climate change movement has at their disposal to show a decline involves the falsification of population data. This falsification is abetted by international organizations sympathetic to the climate change agenda. For example, the Polar Bear Specialist Group (PBSG).[27] which is a subdivision of the International Union for the Conservation of Nature provides the 'official' polar bear census to organizations such as the United Nations. Fortunately, the public has access to *all* polar bear census data, rather than the more limited data set that the PBSG has chosen to report (Fig. 7.4).

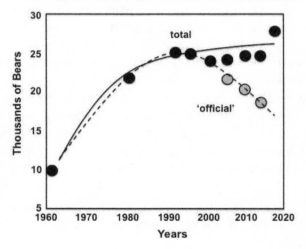

Fig. 7.4 Curves showing 'official' (dashed) and actual (solid curve) records of polar bear populations since 1960. Bear populations could have been as low as 5,000 individuals in 1960 due to overhunting. The count for the most recent census in 2016 indicates a further population increase to 28,000 bears. From data in Ref. 28.

The 'official' report (dashed curve) shows that polar bear numbers have been dropping at an alarming rate. In fact, *their curve indicates that polar bears should be completely extinct in 50 years.* Hmm. *Then why aren't polar bears classified as extremely*

124

endangered or at the very least endangered rather than as simply vulnerable? The bottom 'unofficial' curve incorporating all reported data clears up this discrepancy.[28] It shows that between 1990 and 2013, the polar bear population was stable at 25,000. The most recent 2017 census shows that the population has increased to 28,000. This means that polar bears may not even be 'vulnerable.'

How can the 'official' and 'unofficial' polar bear census numbers be so different? The answer is that PBSG needed to create a population curve based on the premise that polar bear numbers have been dropping since 2000 regardless of actual census reports. Any census numbers that they didn't happen to like were simply discarded. For example, the Russians obviously don't know how to count polar bears. When the Russians reported that 2000 bears had been sighted in the Chukchi Sea, and between 800 and 1200 were seen in the Laptev Sea, these sightings were deemed to be 'inaccurate.' Rather than providing error bars for the perceived uncertainty, the number used by the PBSG from both regions was arbitrarily dropped to zero. The Danes in East Greenland reported 2000 sightings. These must also have been 'inaccurate,' as the total entered by PBSG results was again zero. Even the Canadian report of 200 bears in the Queen Elizabeth Islands was dropped to zero. By this means, the polar bear population magically dropped by over 5,000 animals . . . at least on paper. Climate change advocates may have to resort to violating the current hunting ban in order to bring actual polar bear populations into line with their 'politically correct' agenda.

Actual Changes in Walrus Populations

As the narrative regarding polar bears is losing traction, the climate change movement has been trying to shift attention to the walrus as the species most endangered by global warming. Environmentalists are expressing outrage that the current U.S. Fish and Wildlife Service has not placed the walrus on the endangered species list.[29] They claim that walrus populations are dropping at a catastrophic rate because: 1) without ice to rest on, walrus can no longer reach food supplies on the ocean floor, resulting in starvation, and 2) walrus have been forced ashore in unprecedented numbers because all of the pack ice is gone. Young animals (up to

125

10,000) are being trampled to death on shore, and females with babies are suffering so much that reproductive rates are plummeting.

It is true that there are areas of walrus habitat formerly covered with ice that are now sometimes ice-free. Have these habitat changes *actually* wiped out walrus populations? Do walrus even need ice to survive? The geologic record says no. The walrus evolved around 14 million years ago[30] when global temperatures were 5-10°F higher than they are today and their tropical to subtropical habitats contained little if any ice. Since Ice Ages started 2.5 million years ago, there have been 24 major Ice Age cycles. The high temperature end of each cycle exceeded modern temperatures and ice levels were lower than they are today. Along with the polar bears, the walrus survived all of these cycles unscathed.

What has *actually* happened to walrus populations in modern times?[31] Census numbers show that more walruses are hauling up on land mainly because there are more walrus to haul up. If they are currently 'suffering,' it is due to overcrowding. Between 1900 and 1950, walrus populations dropped to 50,000 individuals. This drop was not caused by global warming but by overhunting. Since then:

*Fossil fuel emissions have increased by more than a factor of ten while the walrus population has **increased** to between 250,000 and 300,000 individuals (by a factor of 5 to 6).*

Based on population alone, one might be forced to conclude that global warming has been causing the walrus population to explode rather than decline. (In fact, much of the increase is due to hunting restrictions). In addition, studies show that: 1) Walrus food supplies have been increasing rather than decreasing,[32] This is because pack ice blocks access to both light and walrus above shallow continental shelves. Partial removal of this ice: a) allows more sunlight into polar waters, b) increases the photosynthesis responsible for creating food supplies, and c) makes it easier for walrus to reach those supplies. 2) While it is true that more walrus calves are being crushed due to overcrowding, it has been clearly shown that the number of calves per cow and the survival rate of those calves have both increased in modern times.[31] In other words:

Walrus are barely a threatened species, let alone one that is in danger of extinction.

Shaye Wolf, climate science director for the Center for Biological Diversity, stated that the U.S. Fish and Wildlife Service is "anti-science," "anti-wildlife," "denies the reality of climate change," and is guilty of "misrepresentation of science to reach a predetermined conclusion" for not declaring that the walrus is an endangered species.[29] Hmm. Who is actually reaching a predetermined conclusion here? Perhaps people like Wolf are the ones who need to look at hard facts and hard science instead of making emotional outbursts that have no basis in reality.

Oceanic 'Acidity'

According to climate change advocates, increasing temperatures are not the only problem, nor is the destruction of life confined to Arctic habitats. Progressives are now claiming that carbon dioxide emissions are making the oceans so acidic that *all* marine life is threatened. Not to be outdone by Al Gore's movie *An Inconvenient Truth,* the National Resources Defense Council (NRDC, a group of lawyers) produced a documentary[33] in 2009 starring Sigourney Weaver entitled *Acid Test: The Global Challenge of Ocean Acidification.* This documentary apparently qualified actress Weaver as a world-expert on oceanic chemistry and biology, as she was called upon to give testimony before the United States Senate on a hearing regarding ocean acidification on April 22, 2010.[34]

The climate change narrative presented in the movie is based on the following chain of logic: 1) When carbon dioxide dissolves in water, it forms carbonic acid (H_2CO_3)(see Appendix 5). 2) Many marine organisms cannot tolerate exposure to acids. 3) Organisms that are most threatened have shells made of calcium carbonate because calcium carbonate dissolves in acid (see Appendix 5). 4) As many species of plankton, as well as corals, oysters, and clams, have calcium carbonate shells, the acidic conditions that fossil fuel emissions are inducing will soon prevent these species from forming and preserving their shells. 5) Much of the marine food chain is based on shell-forming organisms. Their impending extinction due to the use of fossil fuels will wipe out untold additional species of marine life.

127

How much of this is true? Are fossil fuel emissions really converting the ocean into a vast pit of acid? Is all marine life, starting with shellfish, in danger of extinction?

The geologic record says no. Life evolved in the ocean over two billion years ago[35] when carbon dioxide levels were *500 times* higher than they are today. Shellfish evolved in the ocean 500 million years ago[36] when carbon dioxide levels were 150 times (15,000%!) higher than they are today. Yet you are supposed to believe that shellfish and all marine life are threatened with extinction because carbon dioxide levels are 1.3 times or 30% higher than they were 130 years ago. If a 30% increase in carbon dioxide is making the ocean so acidic that shellfish can no longer make shells, how did ancient shellfish produce shells in an oceanic environment containing *150 times* more carbonic acid than is present in marine environments today? The obvious answer is that the carbon dioxide from fossil fuel emissions is not converting the oceans into a hostile acidic environment either now or in the future. In order to understand why this is true, one needs to have an understanding of what makes water either acidic or basic, and also what actually happens to carbon dioxide when it dissolves in water. As most readers are not chemists, a brief primer on acids, bases, pH, and the aqueous chemistry of carbon dioxide is provided in Appendix 5. The conclusions drawn from this chemistry are provided below.

Pure water is neutral, which means that it contains equal concentrations of protons (H^+) and hydroxide ions (OH^-) (10^{-7} moles per liter (M) of each, or one proton and hydroxide ion for every 550 million water molecules).[37] Acidic solutions contain more protons than hydroxide ions, while basic solutions contain an excess of hydroxide ions. The concentration of protons or hydroxide ions determines the acidity or basicity of a given solution, which is typically expressed in terms of *solution pH*. Pure neutral water has a pH of 7 corresponding to $[H^+] = 10^{-7}$ M. Solutions having a pH below 7 are acidic, while those with a pH above 7 are basic. As pH is expressed as a power of ten, a pH 6 solution contains ten times more $[H^+]$ (i.e. 10^{-6} M) than a neutral solution, and is thus ten times more acidic than neutral water. A strong acid such as a 1 M solution of hydrochloric acid (pH 0) contains ten million times more H^+ than neutral water.

When you come in from the rain, is it because you are afraid of getting wet or because you are afraid of being doused with acid as climate change advocates claim? Rainwater is pure water into which atmospheric carbon dioxide has dissolved. Some of the dissolved CO_2 reacts with water to form the 'dreaded' carbonic acid (H_2CO_3). Some of the carbonic acid dissociates to introduce H^+ and the bicarbonate ion (HCO_3^-) into the water, which means that rain water is in fact acidic. Just how acidic is rainwater? How acidic might rainwater become due to future fossil fuel emissions? The answer is: not acidic enough to worry about.

The net effect of CO_2 dissolution, carbonic acid formation, and carbonic acid dissociation (see Appendix 5) generates a proton concentration of 2.3×10^{-6} M and a pH of 5.63. This is only 23 times the proton concentration in pure water or *425,000 times less* than that in a strong acid such as 1 M HCl. Even if all known fossil fuel reserves on Earth were instantly burned to raise atmospheric CO_2 levels to 1100 ppmv (which is impossible) and even if all of this CO_2 remained in the air rather than dissolving in the oceans (which is also impossible, see Chapter 4), the maximum acidity that could be achieved in rainwater by burning all fossil fuels would be a concentration of 3.8×10^{-6} M and a pH of 5.43 (i.e. less than twice the number of protons in existing rainwater).

Of course much more carbon dioxide dissolves into the oceans than in rainwater (see Chapter 4). Does this mean that carbon dioxide from fossil fuel emissions is transforming the oceans into a massive vat of acid? The answer is no. In fact, fossil fuel emissions have less of an impact on the pH of the oceans than they have on rainwater. Seawater is not pure water like the rain, but contains many dissolved salts and other inorganic substances.[38] The net result is that seawater is a basic solution with a pH of 8.1-8.2. In fact, seawater is sufficiently basic that most of the dissolved carbon dioxide is not present as carbonic acid, but as dissolved sodium bicarbonate. Some of the bicarbonate ions lose a second proton to form significant concentrations of the carbonate ions that are present in seashells.

Anyone who has suffered from acid indigestion knows that sodium bicarbonate and sodium carbonate solutions are pH buffers that neutralize 'excess stomach acid' (see Appendix 5). This is why CO_2 additions are less effective in dropping the pH of the oceans than that of rainwater. Geochemists have developed computer programs[39] that can simultaneously solve all of the equations involving equilibrium constants between all species involving dissolved carbon dioxide, including carbonic acid, the bicarbonate ion, the carbonate ion, and calcium carbonate (see Appendix 5). These calculations show that the maximum decrease in pH that could occur by burning all fossil fuels on Earth would be by 0.1 units to a pH of 8.0. In other words:

The oceans are not acidic now. There is not enough fossil fuel on Earth to make the oceans acidic in the future. The oceans were not even acidic over 500 million years ago when shellfish first evolved and when atmospheric carbon dioxide concentrations were 150 times higher than they are today.

Is a pH drop by 0.22 units sufficient to threaten all aquatic life in fresh water (derived from rainwater) with extinction? Is a pH drop by 0.1 units sufficient to threaten all marine life in the oceans? Is either pH drop sufficient to cause the shells of all marine creatures to dissolve? Anyone who owns an aquarium knows that the answer to all of these questions is no.[40] Most aquatic plants and animals are adapted to live in a range of pH of between 0.5 and 1 units. Changes in pH within this range do not have adverse effects unless the pH change is instantaneous. Regarding shells, it is true that: 1) shells are composed of calcium carbonate, and 2) calcium carbonate is more soluble and dissolves to a greater extent as the pH is lowered (see Appendix 5). However, shell-forming organisms can form and preserve their shells at a much lower pH than will ever be seen in the oceans. In fact, fresh-water clams and oysters thrive in environments having a pH as low as 5.5. To summarize this section:

Life in the oceans is not threatened by carbon dioxide either now or in the future. The exact opposite is true. Most oceanic food chains depend on photosynthesis based on carbon dioxide. The idea that carbon dioxide will destroy all marine life on Earth is total nonsense.

Then Senate Majority Leader George J. Mitchell made statements in 1991 regarding the potential impact of Global Warming on humanity, including:[41] "Climate extremes would trigger meteorological chaos – raging hurricanes such as we have never seen, capable of killing millions of people . . . and profound drought that could drive Africa and the entire Indian subcontinent over the edge into mass starvation." "Even if we could stop all greenhouse gas emissions today, [by the end of the twenty-first century] it would be warmer than it has been for the past two million years. Unchecked, it would match nuclear war in its potential for devastation." Pretty strong stuff!

Unfortunately for the climate change agenda, none of the catastrophic weather patterns described by the Senator have any basis in fact. For starters, global temperatures were higher than today as recently as the 1930s. There is no need to go back two million years to see any effects associated with warmer temperatures (see Chapters 3, 6, and 9). In fact, as outlined below, none of the disastrous consequences for humanity claimed by the climate change movement are validated by the bulk of human history. The human race has been exposed to hundreds of climate cycles during which average temperatures were both warmer and colder than they are today. Was humanity pushed to the brink of extinction by raging hurricanes and droughts every time the climate warmed? In fact, a study of historical, geological, and archeological records reveals that if anything, the opposite is true:

Global cooling represents a much greater threat to humanity than global warming.

The earliest known human fossils are 2.5 million years old.[42] This age corresponds to the time when the current era of recurring Ice Ages commenced. Since then, humans have experienced 24 Ice Age cycles. During each Ice Age cycle, the average temperature has fluctuated by 6°C (14°F)(Chapter 3). Sea levels have fluctuated by up to 130 meters (425 feet). None of these cycles wiped out humanity. The only known prehistoric event that comes close to matching Senator Mitchell's 'nuclear war devastation' was the massive volcanic eruption of Mount Toba 74,000 years ago.[43] This mega-volcano caused a short-lived global winter (i.e. global cooling!) that fossil and geologic records indicate could have reduced global human populations from 100,000 down to as few as 3,000 to 10,000 individuals.[43]

All human civilizations have arisen during the period of Global Warming that has existed since the end of the last Ice Age. Since the last Ice Age, there have been five extended periods when average temperatures were as warm or warmer than they are today, as well as five extended periods of global cooling (Fig. 7.5).[44]

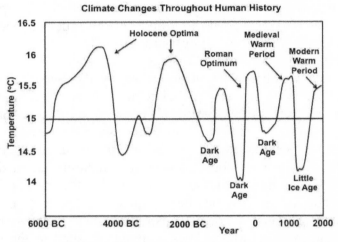

Fig. 7.5 A time line showing climate variations that have occurred during the past 10,000 years of human history (adapted from Ref. 44). Human health and major cultural advances have coincided with warm periods such as the two Holocene Climate Optima, the Roman Climate Optimum, and the Medieval Warm Period. Cold periods tend to coincide of eras of famine, pestilence, and social unrest called Dark Ages.

How did humanity fare during each of these periods? The welfare of humanity is highly dependent on the availability of ample food and water. Terrestrial food supplies are enhanced by warmer climates. Global water supplies are also enhanced because more water evaporates from warm oceans than cold oceans, resulting in higher levels atmospheric moisture and terrestrial precipitation. While arid regions still exist, archeological and historical records clearly show that a larger fraction of the Earth experiences wetter conditions during warm rather than cold eras.

Clear metrics that tie the welfare of humanity to warmer climates include:[41] 1) relative human populations have surged during warm periods and ebbed or stagnated during cold spells, and 2) skeletal remains show that humans grow larger, live longer, and show less evidence of disease during warm eras than cold eras. None of this should come as a big surprise. History clearly shows that humanity has not been devastated or threatened with extinction every time the climate gets warm. Specific events associated with the warm and cold eras only serve to confirm that if humans could indeed control the Earth's climate (which they can't), then people should deliberately try to make the climate warmer.

The two warmest periods that the Earth has experienced since the last Ice Age are called the Holocene Climate Optima (Fig. 7.5). In both optima, global temperatures were substantially warmer than they are today. Prior to the first optimum, humans lived as nomadic hunter-gatherers who survived exclusively on food sources they could find. The first Holocene Climate Optimum from around 6000-4000 B.C. coincides with the most important advance in human history: the domestication of plants and animals for agriculture.[45] The major global proliferation of advanced civilizations based on agriculture occurred during the second Holocene Climate Optimum from between 3000-1300 B.C. These periods are called *optima* because their benign warmer climates were 'optimum' for growing crops and sustaining human health.

The subsequent cold Submycenaean period between 1300-900 B.C. hit regions such as the Mediterranean hard. This period was characterized by such extensive depopulation and collapse of centralized political power that it has been called one of Europe's first Dark Ages.[46] However, Greek city-states recovered and blossomed during the following warm period. After a cooling period, the next warm period from roughly 300 B.C. to 300 A.D. is called the Roman Climate Optimum. The mighty Roman Empire flourished during this period, reaching its maximum size and prosperity around 200 A.D.[47]

The collapse of the Roman Empire took place during the following period of global cooling lasting from 300-1000 A.D. Much of this cooling period coincided the most famous of Europe's Dark Ages[48] from 500-1000 A.D. The Dark Ages was a period of extreme poverty, starvation, and more epidemics than in any other historical era. Europe came out of the Dark Ages and into a new era of prosperity and cultural development during the following Medieval Warm Period (see Chapter 3) from around 1100-1350 A.D. Temperatures during this rebirth or Renaissance[49] were once again warmer than they are today.

The subsequent cooling period, called the Little Ice Age, saw the end of the Renaissance. This era of cold temperatures, lasting from roughly 1400-1850, was one of poor growing conditions, famines, riots, and social unrest across much of Europe.[41] Severe famines killed millions of people in 1527-1529, 1590-1597, the 1640s, 1690-1700, and 1725. The Great Famine in 1740-1741 killed almost one million people in Ireland alone,[50] or 38% of the entire population. The famine in 1789 triggered the French Revolution.[51] The Irish Potato Famine[52] of 1845-1852, which killed another million people in Ireland, was the last major famine prior to the modern era of warmer temperatures.

Is it just a coincidence that all of the warm periods described above were accompanied by eras of relative prosperity and cultural development? Is it just a coincidence that all of the cold periods were associated with eras of crop failures, famines, poverty, and pestilence? Perhaps. However, the entire history of humanity suggests that people should have nothing to fear from any warming that the Earth is currently experiencing, let alone unsubstantiated warming events associated with the combustion of fossil fuels.

Summary

Carbon dioxide is essential to all life on Earth. This gas forms the basis for photosynthesis, which provides the energy and organic compounds that support the entire food chain. Carbon dioxide is not a toxic pollutant. While the explosion of human populations has led to the extinction of many plant and animal species, there is no proof that a single species is currently being driven to extinction due to man-made global warming. The geologic record clearly shows that global cooling poses a more serious threat to life on Earth than global warming. If current extinctions are to be blamed on climate change, the blame is associated with the fact that currently climactic conditions are ideal for the proliferation of humanity.

Chapter 8: Renewable Energy

Myth: 'Green' renewable energy sources managed by the government represent the only real solution to the global warming problem. Solar power, wind power, and biofuels will save our planet from the ravages of fossil fuels.

According to climate change advocates, there is only one possible solution to the problem of man-made global warming: eliminate the use of all fossil fuels. How is this to be done? The climate change community insists that climate-destroying fossil fuel pollution can be terminated forever if the world switches their energy production template from the use of fossil fuels to 'green' technologies that produce no pollution and are environmentally friendly. The three renewable energy supplies that are most strongly touted include: 1) solar energy, 2) wind energy, and 3) 'all-natural' biofuels. Unfortunately, the real motivation of the climate change blueprint is to shift control of all energy supplies from oil companies and other 'evil' corporate entities to 'benign' socialist governments. The allure of renewable energy sources is that they provide a mechanism for gradually transferring the control of energy from the private sector to Big Government without the public even being aware that the transfer is happening.

Renewable energy is expensive and inefficient. Most renewable energy sources only become economically viable when they receive massive government subsidies (see specific examples below). At a minimum, these subsidies allow governments to assert control over the renewable energy sector. Unfortunately, these subsidies are also being used to try to put fossil fuel companies out of business. This works because *the subsidies and tax breaks afforded to renewable energy do not appear as direct costs to the consumer but are hidden within a tax burden that is shared by everyone.* Most people have no idea how much they are *really* paying to support the renewable energy agenda. Governments can set the tax burden for subsidies at any level they wish. This allows governments to set the

135

prices charged to consumers at any level they wish. Governments have the power to make it *appear* that renewable energy sources no longer cost any more than fossil fuels. This strategy is being used to convince the public that there is no longer any reason to use 'planet destroying' fossil fuels, as 'green' renewable energy is just as economical. The unpleasant surprise for consumers will come once governments succeed in eliminating all fossil fuel competition. Then the need for government subsidies will disappear and consumers will be forced to pay the entire cost while suffering limited access to energy associated with renewable sources controlled by Big Government.

How plentiful and economical *are* renewable energy sources? To what extent can renewable energy actually meet the energy needs of advanced industrialized societies such as the United States? Below, the energy portfolio of the United States is provided to put the potential role of renewable energy sources into perspective. This is followed by a discussion of specific renewable energy sources, and what the prospects really are for sustaining the future energy needs of the industrialized world.

Energy Use in America

Consider how the energy needs of the United States are currently being met. The United States uses an enormous amount of energy.[1] As of 2015, the total energy consumed in America was the equivalent of 97.5 quads (Fig. 8.1), where a quad stands for one quadrillion (10^{15}) British thermal units (Btu)(see Appendix 3). This energy is equivalent to:
- 17,000,000,000 (17 billion) barrels of oil
- 28 trillion kilowatt-hours per year of electricity (Appendix 3)
- the continuous operation of 1.2 trillion 100-watt light bulbs
   (4,000 light bulbs for every person in America)

Of this energy, 38% of consumption involves electricity, 29% is used for transportation, while much of the remaining 33% involves on site power generation by industry and businesses. Many Americans might be surprised to learn that only 11% of the country's energy use involves providing power to residential neighborhoods.

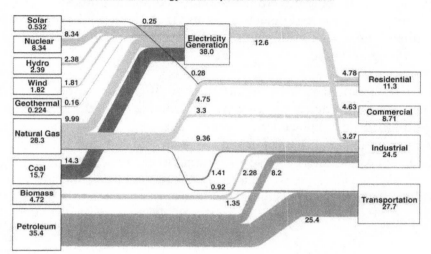

Fig. 8.1 A simplified version of a Department of Energy chart showing the flow of the total amount of energy in quads (see text) throughout the United States economy.[1] Sources of energy are listed on the left, while consumers of the energy are listed on the right.

Fossil fuels (petroleum, natural gas, and coal) represent almost 82% of all energy used in America. Nuclear power provides 9%, and hydroelectric power represents 2.5%. Of the renewable energy sources preferred by environmentalists, biofuels such as ethanol in gasoline represent 5%, wind power is 2%, and solar energy is only 0.5% (half a percent) of our energy total. Over 92% of the energy used for transportation is provided by petroleum. Natural gas and coal generate 64% of our electricity. The on-site generation of energy used by heavy industry is almost entirely based on natural gas and petroleum.

The climate change blueprint for the energy future of the United States extends beyond the use of fossil fuels to include the following elements:

- Close all power plants burning fossil fuels, starting with coal.

- Shut down all nuclear power plants, as anything radioactive is inherently evil.
- Force a shift from gasoline-powered vehicles to electric vehicles, with the ideal being government-controlled public transport.
- Close as many hydroelectric plants as possible and prevent the United States from building any more, as dams block access to fish spawning grounds.

This means that 95% of the massive 30 trillion kilowatt-hours per year that currently power our society would have to be eliminated and replaced by 'green' renewable energy sources. Is this remotely possible? If fossil fuels are so evil and destructive, and renewable energy is so clean and wonderful, why hasn't the United States and the world already abandoned fossil fuels in favor of a renewable energy economy?

Solar Energy

Consider solar energy, which currently represents only half a percent of the total energy used in America. There are multiple reasons why this percentage has not been higher, including:

Solar power is inefficient. Under ideal conditions (noon at the Equator), the maximum amount of sunlight hitting one square meter of the Earth's surface (roughly a square yard) delivers 1000 watts (W) of power[2] or 1 kilowatt/m^2. The average for a normal 12-hour day due to shifting sun angles drops the maximum to 600 W/m^2. Commercial photovoltaic collectors can only harvest around 15% of that power.[3] This means that a solar collector having a surface area of 1 m^2 gathers 90 W/m^2 under ideal conditions, or about enough energy to power a single 100-watt light bulb. The sun doesn't shine at night, dropping the maximum theoretical harvesting for 24 hours down to 45 W/m^2. Collectors don't occupy all of the space in a solar plant. This drops the maximum power rating under ideal conditions for a typical solar facility (i.e. the number that is always reported) down to around 25 W/m^2. Unfortunately, conditions are rarely ideal, so the maximum rating is never realized. Is it cloudy? Then collection drops by another factor of three. Is it smoky or dusty? Is the collector dirty? Collection

138

could drop all the way down to zero. The percentage of the maximum rating that is delivered in practice is called the *capacity factor*. The capacity factor averaged over all U.S. solar energy facilities is 20-30%.[4] *This means that the average output of a typical solar power facility is really only 5-7.5 W/m^2.*

Photovoltaic chips used in most power plants consume more energy than they collect. This is because the steps required to convert raw quartz (SiO_2) feedstock into the high-purity doped silicon wafers used for light collection are extremely energy intensive. In terms of kilowatt-hours (kWh) per kilogram of silicon (Si) produced, these steps include: 1) reacting quartz with carbon to produce Si (13 kWh, releasing CO_2 as a bi-product), 2) conversion of silicon into trichlorosilane gas for purification (50 kWh), 3) reaction of trichlorosilane back into polysilicon (250 kWh), 4) high-temperature processing to transform polysilicon into a single crystal ingot (250 kWh), 5) producing Si wafers from the ingot (240 kWh), and finally 6) fabricating semiconducting chips from the wafers in expensive clean-room facilities (2130 kWh). The grand total of 2930 kWh per kilogram of silicon is equivalent to 3370 kWh per square meter (m^2) of collector material produced. Given the fact that the collector efficiency in a power plant setting is only 7 W/m^2, a photovoltaic collector needs to harvest solar energy for over 50 years (exceeding the expected service life) in order to reach the break-even point where the energy harvested exceeds the energy that was used to produce the collector in the first place.

Solar energy can't be turned on and off to meet shifts in energy demands. The sun shines during the day, while power use peaks in the morning and at night. Much less energy is collected in the winter than the summer due to shorter days and changes in solar angles. There are two potential solutions to the intermittent nature of solar energy. The current solution is to use conventional (i.e. fossil fuel) sources during off-peak hours. This means that existing fossil fuel plants have not been eliminated and continue to generate CO_2. However, consumers are now paying for two power plants instead of just one. The future utopian solution involves eliminating fossil fuel plants altogether. However, for this option, a substantial fraction of the energy collected during the day has to be stored in massive, expensive, and inefficient battery facilities for use at night. A typical lead-acid car battery has a storage capacity of around 1 kilowatt-hour.[6] A total replacement of fossil fuels by solar energy would require the equivalent of 15 trillion of such batteries. More efficient lithium batteries are so expensive that they would more than double the already high cost of generating electric power via solar energy (see Wind Energy below).

Solar power plants must be built where there is plentiful and continuous sunshine like the middle of the desert Southwest rather than at the urban locations where the energy is needed such as the Northeastern corridor.

Photovoltaic Solar Resource of the United States

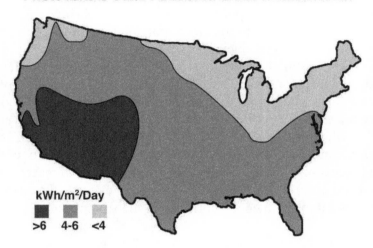

kWh/m²/Day
>6   4-6   <4

Fig. 8.2 A map of the United States showing the amount of sunlight that is available for harvesting by photovoltaic devices by geographic region. The meaning of the colors is provided by the bar at the bottom in units of kilowatt-hours per square meter per day. Adapted from data in Ref. 7.

Of course, solar facilities can be built almost anywhere. However, Figure 8.2 shows that the solar energy available to be harvested in the Northeast is half of what it is in the Southwest.[7] This halves the efficiency and doubles the cost per kilowatt-hour of solar-powered electricity generated in the Northeast. New and expensive transmission lines must be built to connect remote solar power facilities to end-users. As the costs and resistive heating losses associated with transmission lines scale with distance, the costs associated with the transmission of solar energy will start to skyrocket.

Because solar panels are so inefficient, the land area required to generate significant amounts of energy is huge[8]. For example, the largest solar power plant in America (Cal-Renew-1) is a 5 million watt (5 MW) maximum power rated facility that occupies 50 acres of what used to be prime farmland. Maximum power generation on the site amounts to 100 kW/acre (or 25 W/m^2). When multiplied by the most generous solar capacity factor of 30% (see above), the actual power generation at the site is 7.5 W/m^2. A 1000 MW plant would require a land area of 51 square miles, exceeding the total area of San Francisco (Fig. 8.3).

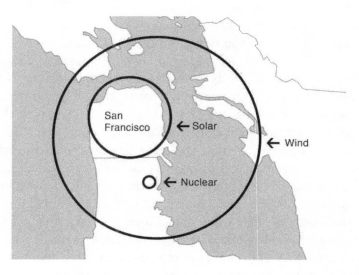

Fig. 8.3 A map of the San Francisco Bay area in California used to illustrate the land area required by different energy technologies to generate 1,000 megawatts (1 billion watts) of energy. Circles confine the land area required by nuclear, solar, and wind power plants.

In contrast, the Diablo Canyon nuclear power plant (2025 MW of actual power on 900 acres), consumes 80 times less land[7] (0.624 mi^2 per 1000 MW). Where is all of the land needed to replace fossil fuels with solar energy going to come from? What land resources are to be sacrificed in the name of 'green' power? The sober truth is that there is not enough land available in the United States to replace fossil fuels as a source of electricity, let alone enough to replace fossil fuels for transportation and on-site energy needs (see Actual Prospects below).

Solar energy is way too expensive for most countries or individuals to afford. Over 1.5 billion people or 15-20% of the world's population live without any electricity at all.[9] Although coal is vilified for producing 33% of the world's carbon emissions, the use of coal continues to rise to supply 40% of the world's energy.[10] This is because at a real cost of only 7 cents per kilowatt-hour, it is the cheapest form of readily available energy after natural gas at 6 cents. How expensive is solar energy? The costs reported for solar installations have dropped dramatically in recent years to as low as 16 cents per kilowatt-hour. However, *claims regarding affordability ignore the massive government subsidies of 24 cents per kilowatt-hour that are being pumped into solar energy.* An additional 5 cents needs to be added to account for the costs to maintain and fuel backup power and long-range transmission capabilities. This means that the *real* cost of solar energy[11] is 45 cents per kilowatt-hour, which is seven times more expensive than natural gas. In homes, system and installation costs are substantial, with systems providing partial (5 kW or 50%) electrical coverage for a 2,000 square foot home costing an average of $40,000 even after large ($20,000 per home) government subsidies. Few Americans can afford to pay this much to 'save the planet,' let alone people living in poorer developing nations (see Consequences below). Even with enormous government subsidies, the power bills charged to consumers are increasing at an alarming rate to help pay for new solar facilities.

Wind Energy

Wind energy currently provides the United States with four times the amount of energy that solar technologies do. Does this mean that wind power can grow to a sufficient extent to replace fossil fuels? The answer is no. The growth of wind power is hampered by limitations including:

1) Wind power is intermittent and inefficient. The turbines in a wind farm can only harvest energy and convert it into electricity when the wind is blowing. If the wind isn't blowing at all, no energy is harvested. If the wind is blowing too hard, the facility is shut down to inhibit damage to wind turbines. As with solar collectors, wind turbines are rated relative to the maximum power that they can generate. As with solar, the actual energy produced is determined by the product of the maximum power rating times the capacity factor, which is the percentage of the maximum rating that is actually harvested. The capacity factors for current wind farms range from 30-40%.[4,12] However, the average capacity factor will continue to drop as the most ideal sites for wind energy are occupied.

2) Batteries and other energy back-up systems are expensive. As a specific example,[13] in November of 2017, Tesla Chief Executive Elon Musk held a press conference to announce that he had delivered the world's largest battery system to back up the new Hornsdale wind farm being built in South Australia. He claimed that this battery system will be capable of storing enough electricity from the wind farm to power 30,000 homes for an hour. While this battery system represents an impressive technological accomplishment, the cost of the system has been glossed over. The Hornsdale wind farm is projected to cost $25 million. A single Tesla battery system is estimated to cost $50 million, or twice the cost of the wind farm. However, given the capacity factor of a typical wind farm, the facility will probably need to store electricity for two thirds of the time, or 16 hours (not just one hour!) a day. To supply 30,000 homes with electricity for a full 24 hours, 16 Tesla battery systems would be needed, for a total cost of $400 million, or over thirty times the cost of the wind farm itself.

3) Siting of wind facilities is problematical. To generate significant amounts of wind energy, wind power facilities need to be located in areas where there is steady wind most of the time. Is it always windy where you live? If so, you occupy one of the fortunate areas where wind energy can provide a continuous, reliable source of electricity. A map produced by the National

Renewable Energy Laboratory of the Department of Energy[14] (Fig. 8.4) shows that areas having the highest potential for wind energy are along the west coast of the United States and in a strip through the Midwest from north Texas to North Dakota.

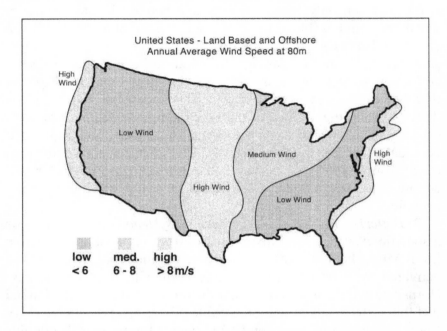

Fig. 8.4 A simplified version of a map prepared by the National Renewable Energy Laboratory (NREL) showing the relative amount of wind energy available for harvesting in different regions of the country.[14] Ranges in wind speeds are given in meters per second.

Roughly 75% of the country has only half of the wind potential of these areas, which translates into a 50% decrease in wind farm energy production. Offshore coastal areas have a higher wind potential but are at least three times more expensive to develop. In addition, given commercial, recreational, and environmental concerns, offshore sites are generally unavailable.

4) Wind power is extremely land intensive. One of the biggest problems regarding wind energy is the large amount of land that it consumes (Fig. 8.2). Wind turbines are large objects. In addition, turbines within wind farms must be spaced sufficiently far apart that they do not interfere with wind collection at neighboring turbines. The federal government's National Renewable Energy

Laboratory (NREL) has reported the most definitive estimate of total land use for wind farms.[14, 15] Data collected for 172 of the largest wind farms indicate that their total capacity of 25,000 MW is housed on 8,800 square miles. Using a generous capacity factor for these farms of 40%, the actual power delivered is 10,000 MW. This means that a wind farm that actually *delivers* 1,000 MW requires a land area of 880 square miles. Even if one assumes that all wind turbines are operating at their maximum capacity 100% of the time, a land area of 350 square miles would be required corresponding to the circle in Fig. 8.2. *The land area required for wind power is over six times greater than that of a solar power plant and 530 times the area occupied by the Diablo Canyon nuclear power plant.*

5) *Wind power is expensive.* Renewable energy companies, environmentalists, and government agencies such as the Energy Information Administration (EIA) report that wind power only costs 8 cents per kilowatt-hour.[16] However, such claims are based on assumptions including: a) the average lifetime of a wind turbine is projected to be 30 years or the same as a conventional power plant, b) the costs of backup power for when the wind doesn't blow are not included, c) wind farms are close to end users to minimize transmission losses, and most importantly d) *government subsidies and tax breaks are not considered in the cost estimates.* The reality of the situation[17] is that: a) Long-term experience in the United Kingdom shows that realistic turbine lifetimes, representing 80% of total costs, are only 15 rather than the claimed 30 years, adding 7 cents to the total. b) Total costs for backup power add at least 2.3 cents if the backup is natural gas. c) Added transmission costs are estimated at 2.7 cents. *The sum of all current government subsidies and tax breaks adds a whopping 23 cents*, making the grand total 43 cents per kilowatt-hour. This cost is equal to that of solar energy or 7 times that of natural gas. Can you afford to have your energy bills increase by a factor of seven? Can U.S. industries afford such increases?

6) *Wind power is not environmentally friendly, generating noise pollution and killing wildlife.* Environmentalists want wind farms to be everywhere . . . except in their neighborhood. Wind turbines are large and ugly and spoil the appearance of the landscape. The noise generated by a wind turbine has been likened to that of a helicopter.

Noise pollution makes human habitation oppressive even a quarter of a mile away. However, the main concern expressed by environmentalists involves the number of birds and bats that wind turbines kill every year. It is estimated that 8,200 birds, including 633 raptors, were killed at the Altamont Pass Wind Farm alone between 2008 and 2009.[18] The Audubon Society estimates that wind farms kill 330,000 birds in North America every year,[19] making wind the deadliest form of renewable energy.

7) *Wind power is dangerous to humans*. As of 2017, there have been 111 human fatalities at U.S. wind farms[20] compared with zero at the supposed 'death traps' represented by nuclear power plants. Most of the fatalities have been caused by catastrophic failures of massive turbine blades. Lawsuits are springing up all across the country due to the harm wind farms inflict on wildlife, the quality of life, and property values. Several governments, including those in Finland, Bavaria, and Scotland are currently proposing legislation to insure that no wind-farms will be allowed within two kilometers of any housing. The Finnish Ministry of Health issued the statement that: "The actors of development of wind energy should understand that no economic or political objective should prevail over the well-being and health of individuals."

Energy from Biofuels

Biofuels, primarily ethanol in gasoline, provide 5% of our current energy needs. According to environmentalists, biofuels represent the ideal short-term solution for replacing fossil fuels for use in transportation and on-site power generation. They argue that because biofuels come from plants, they are 'all-natural,' 'green,' and non-polluting. They claim that in contrast to fossil fuels, biofuels are a renewable and sustainable because a new biofuel crop can be grown every year. Unfortunately, *once one moves away from emotions, rhetoric, and politics, the push for biofuels defies all logic*. It accomplishes nothing 'positive' and much that is negative other than transferring control of transportation fuels from oil companies to the government. The major losers in this power grab are the American people. Here are just a few reasons why biofuels are not the answer to our energy problems.[21,22]

1) Biofuel production cycles and subsequent combustion create more carbon dioxide than fossil fuels. Biofuels are hydrocarbons similar to the molecules found in gasoline. Burning a biofuel produces carbon dioxide. The CO_2 molecules produced from the combustion of ethanol (C_2H_5OH) are not any 'greener,' 'cleaner,' or more natural than the CO_2 molecules produced from the combustion of propane (C_3H_8). The energy (or heat) produced *per molecule of CO_2 emitted* is essentially identical[23] for biofuels (165 kcal/mole for ethanol) and gasoline (163 kcal/mole for octane). However, in order to make a biofuel, the sugars and starches in corn have to be converted into molecules such as ethanol. This conversion involves the same fermentation processes used to make carbonated alcoholic beverages such as beer. Fermentation releases one molecule of carbon dioxide for every molecule of ethanol produced.[24] The carbon dioxide created during fermentation is called 'biogenic' CO_2. Production of CO_2 by fermentation is not any 'greener,' 'cleaner,' or more natural than CO_2 from fossil fuel. Biogenic CO_2 represents half as much CO_2 as that produced during the combustion of ethanol as a fuel. Even if one ignores the fuel consumed to produce the ethanol (see below), a typical biofuel generates 50% more CO_2 than the gasoline it replaces.

2) Biofuels consume more energy than they produce.[25] The corn from an acre of land produces between 330 and 450 gallons of ethanol. As a gallon of ethanol has only 63% of the energy content of gasoline, 400 gallons of ethanol per acre is equivalent to 250 gallons of gasoline. It is estimated that the farm equipment used to plant, tend, and harvest the corn consumes 140 gallons of fossil fuels. After harvesting, processing steps including fermentation and two to three distillation cycles consume additional energy that is also provided by fossil fuels. The bottom line is that it takes around 100,000 BTU to produce a gallon of ethanol that only provides 77,000 BTU of energy. In other words, *ethanol consumes one third more energy than the fuel produces.* In contrast, only 22,000 BTU are expended to create a gallon of gas that provides 120,000 BTU of energy. In addition, when the CO_2 emissions associated with fossil fuel consumption are added to the 'biogenic' CO_2 contribution, this 'green' biofuel actually generates over twice the amount of carbon dioxide that would be generated by just using fossil fuels in the first place.

3) Biofuel production is extremely land intensive. The total land area in the United States that is available for growing crops is around 320 million acres. The total area currently used for growing corn is around 90 million acres, or 28% of all arable land. At the present time, 40% of the corn crop is used to produce ethanol,[26] which amounts to over 10% of all available farmland. Currently, less than 5% of the transportation fuel consumed annually in the United States (140 billion gallons of gasoline and 143 billion gallons of diesel fuel) is ethanol. Replacing all fossil fuels with ethanol would require twenty times the amount of ethanol that is being produced today, requiring the use of a total land area of 720 million acres. This means that even if the nation abandoned all crops other than corn for ethanol, the country could only produce 40% of the fuel required in order to meet current transportation needs. This analysis ignores the fact mentioned above that it takes more energy to produce ethanol than is recovered via ethanol combustion. In reality, *the energy biofuels produce per unit land area is a negative number.*

*4) Biofuels cost more and in some instances **much** more than gasoline.*[24] Ethanol is the cheapest biofuel that is currently available. The difference in price between ethanol and gasoline is mainly in the cost of the feedstock (i.e. corn vs. oil) as processing and transportation costs are similar and relatively minor. At a price for corn of $4.50/bushel, the feedstock cost for ethanol is $1.80/gallon. However, as it takes 1.5 gallons of ethanol to produce the same energy as one gallon of gas, the cost per equivalent energy of ethanol is $2.70. In comparison, the current cost of a barrel of oil (42 gallons) is $50, leading to a feedstock cost of $1.20/gallon. A barrel of oil produces 31 gallons of fuel (19 gallons of gas plus 12 gallons of fuel oil), making the feedstock price equal to $1.35/gallon. In other words, the cost of ethanol per unit energy is twice that of gasoline. This cost does not include the hidden expenses to taxpayers associated with government subsidies and rebates to corn growers and processers. The Congressional Budget Office reports that taxpayers are currently charged $1.78 per gallon of gas replaced. Counting these government contributions, ethanol costs around $4.50 per gallon, or over three times what gasoline costs.[25] If you think that is expensive, consider the plans that were mandated in the Defense Department under President Obama. The

Defense Department goal was to shift the fuels used in aircraft and military vehicles to 'green' isopropanol-based biofuels provided by Gevo, Inc. *at a cost of $59/gallon* or ten times the cost of standard aviation fuel.[27] These fuels, which were claimed to reduce CO_2 emissions by only 20%, would cost the military $300 billion per year if used to completely replace fossil fuels.

5) *Most cars and trucks can't function using pure ethanol as a fuel.* You've heard about electric cars, and maybe even propane trucks, but have you ever heard of an ethanol car? The answer is no. This is because the standard internal combustion engine used in most cars cannot function on pure ethanol. In fact, a wide range of groups, including all major automobile manufacturers and the Automobile Association of America (AAA) have made strong statements to the effect that even increasing ethanol levels in gasoline from the current 10% ceiling up to only 15% would seriously overheat engines and damage many engine components such as fuel pumps.[28]

6) *Biofuel production competes with food production.*[24] Every acre of land now used for biofuel production was formerly used for food production, mainly as feed for livestock and poultry. Over 40% of U.S. corn production has made the switch from food to fuel since 2005. Other crops that have seen their acreage decrease to produce biofuels include: oats (down 32%), barley (down 46%), sorghum (down 43%), wheat (down 14%), and hay (down 6%). Farmers are only responding to the fact that large government subsidies for ethanol (which you pay for!) have driven corn prices to over $4.50/bushel, making it more profitable for them to grow corn for ethanol than other crops. High corn prices and the scarcity of other food stocks have increased food prices across America by 25% since 2005.

The Electric Car

Billionaire Elon Musk knows that a lot of money can be made in businesses that support popular environmental causes. He has gained the fervent support of the 'green' movement by broadcasting his utopian vision for the future of reducing global warming, establishing a sustainable energy economy, and even establishing a human colony on Mars. He owns the electric car company Tesla,

Inc. His aerospace company SpaceX is developing a 600 mph transportation system called the Hyperloop based on electric propulsion and magnetic levitation of vehicles traveling through evacuated tubes. Tesla cars and Hyperloop vehicles are all to be powered by solar energy provided by his solar energy company SolarCity.

At first glance, Musk's futuristic vision for electric cars and public transportation *looks* like the ideal path for preserving the planet. After all, an electric car has no tailpipe and obviously produces no emissions. Government-subsidized electric buses and trains have been proliferating in recent years to prove that government officials are concerned about our climate. However, people seem to be ignoring two key questions regarding electric transportation:

- Where does the electricity used to power electric cars come from?
- What energy sources are capable of meeting transportation needs in the future?

To evaluate the true potential for electric cars, consider that:

1) Electric cars actually generate almost as much carbon dioxide as gasoline-powered cars. People tend to lose sight of the fact that most of our electricity (65%) comes from power plants that burn fossil fuels. The energy efficiency of a fossil fuel power plant is about the same as that of a car. Carbon dioxide emissions per unit energy are also nearly identical. However, 11% of the electricity generated is consumed in transmission and distribution to electrical outlets. An additional 15% of the electricity is lost in charging and discharging even the most advanced lithium batteries. The net result is that electric cars currently generate 85% of the carbon dioxide emissions of a gasoline-powered car having the same size. Driving a car that is 15% smaller has the same impact as driving an electric car in terms of reducing the CO_2 footprint.

*2) The nation cannot **produce** sufficient electrical energy to replace gasoline.* The current total electrical generation capacity of America is 11.4 trillion kilowatt-hours.[1] The total energy used for transportation is 8.5 trillion kilowatt-hours. Solar and wind currently generate 0.7 trillion kilowatt-hours, or 6% of our electrical and 2.5% of our total energy needs. For the vision of the all-electric transportation future to be realized, the entire transportation total plus the fossil fuel component (65%) of electricity generation must be replaced with solar and wind energy. This would require the construction of new solar and wind farms having a total electrical generation capacity of 16 trillion kilowatt-hours, or 22 times the capacity that is currently in operation. We have neither the economic resources nor the land area to come close to meeting this total.

*3) The nation cannot **distribute** sufficient electricity to replace gasoline.* The electrical power grid of the United States is currently valued at $900 billion.[28] Like much of America's infrastructure, the grid is in dire need of hundreds of billions of dollars in repairs and modernization to replace expensive components such as transformers that are 40-50 years old. Power outages including blackouts have increased by a factor of three since 1984. In fact, the U.S. power grid experiences more blackouts than that of any other developed nation.[28] The largest blackout in U.S. history occurred in 2003 in the New York City area, leaving 50 million people without power. It is estimated that blackouts currently cost businesses over $150 billion/year. What do proponents of electric cars expect this aging grid to accomplish?

Fig. 8.5 A map showing the distribution of people in the United States in the year 2000 (adapted from data in Ref. 29). Each dot on the map represents 7,500 people.

At a minimum, new demand associated with converting all transportation from fossil fuels to electric power will require the grid to increase its current capacity by 72% (28 quadrillion watts, see Fig. 8.1). However, if one examines where the U.S. population resides[30] (Fig. 8.5) relative to where the major hubs of solar and wind power are located (Figs. 8.2 and 8.4), it is easy to see that a total shift to renewable energy could require a grid that is more than twice its current size.

Even if sufficient renewable energy sources could be built to replace gasoline and fossil fuel plants, the costs and technical problems associated with upgrading the power grid to accommodate the huge new demand would be astronomical.

The Consequences of Government Policies Regarding Renewable Energy

What are the projected costs and final outcomes associated with accomplishing nothing in the name of renewable energy and climate change?

Below, several examples are provided to illustrate the challenges faced in government attempts to try to force the conversion from fossil fuels to renewable energy.

Germany

The costs per unit benefit of renewable energy are highlighted by the experience of Germany.[31] Germany has proclaimed its mission is to be the world leader in renewable energy. It is constantly chastising the United States for not doing more to curtail the problem of global warming. Germany has pledged to produce 80% of its energy from renewables by 2050. Germany's environmental minister has estimated that hitting this target will cost another $1.3 trillion over the next twenty years. As the United States consumes seven times the energy of Germany, the equivalent cost for America would be over $9 trillion (nearly half of the national debt).

How close has Germany come to actually meeting their goal of a renewable energy economy? As of 2016, almost all (81%) of Germany's energy needs were still being provided by fossil fuels

(34% oil, 24% coal, and 23% natural gas). Almost all of these fossil fuels (91%) were imported. Only 2.1% of Germany's energy is provided by wind, and even less (1.2%) is provided by solar. Carbon dioxide emissions are down by a whopping 0.1% (1/1000). What about the costs? Germans have seen a 47% increase in their electricity bills since 2006 even though wind and solar provide only 3.3% of their electrical needs. There has also been a ten-fold increase in renewable energy subsidies provided to wind and solar energy concerns.

Germany is paying a steep price in order to have no impact on global warming.

Spain

During his term in office, President Obama pointed to Spain as the country that the United States and the rest of the world should emulate. At that time, Spain was pursuing Europe's most aggressive shift into a renewable energy economy. However, a detailed economic analysis[32] clearly shows that Spain's policies had a negligible impact on CO_2 emissions while causing extensive damage to Spain's economy. Highlights of the 2009 report include:

As in the United States, the real cost for solar energy, which was providing less than 1% of Spain's electricity and was thus having next to no impact on Spain's (let alone the world's!) greenhouse gas emissions, was seven times that of fossil fuels.

The added costs of renewable energy were being paid for out of government subsidies ($36 B) and increased electric bills (up by 31% or $10 B) for a total of $46 B. When scaled to the size of the U.S. economy, their total added costs would be equal to $690 B.

While renewable energy has been touted for creating 'green' jobs, it has been shown that for each 'green' job created, 2.2 jobs were destroyed elsewhere in the economy due to higher energy costs. When scaled to the U.S. economy, job losses from similar policies are projected to range from 6.6 to 11 million jobs. Spain spent $720,000 per 'green' job created, with the high being over $1,250,000 for wind energy jobs.

Spain has paid a steep price to have no impact on global warming.

The United States

Experiences of the United States mirror those of Germany and Spain, including:

1) Global CO$_2$ emissions are essentially unchanged. As discussed throughout this book, carbon dioxide emissions are not destroying the Earth's climate. However, even for those who believe in man-made global warming, it is clear that the aggressive push to replace fossil fuels with solar and wind energy throughout America is having no impact on the problem. At this point in time, the United States generates only 14% of global CO$_2$ emissions[33] behind China's 30%. Of that 14%, only 3% (or 0.4% of the global total) has been replaced by wind and solar energy. This means that all of the sacrifices that Americans have made in the name of 'saving the planet' have lowered CO$_2$ emissions by a whopping 0.4% (less than half a percent or 1/250 of the global total). On a global scale, the push toward renewable CO$_2$-free fuels in the United States has been completely irrelevant.

2) Consumers pay much more for electricity, gasoline, and food. Everyone is losing money in the rush to embrace renewable energy sources that have no impact on a problem that doesn't even exist. A dirty little secret is that technologies have already been developed to remove all CO$_2$ from power plant emissions.[34] These technologies are not used because regeneration of scrubbing columns would consume half of the electricity generated by the plant. The resultant doubling of power rates would put power companies out of business, since neither private nor industrial customers can afford to pay for that. Yet somehow Americans are expected to embrace renewable energy technologies that provide energy at seven times the cost of fossil fuels. Government mandated quotas on solar and wind power production have already increased electricity prices by about the same factor of two required to scrub existing fossil fuel plants. Americans pay for biofuels that are less efficient, cost at least three times more, and generate more CO$_2$ emissions than gasoline. Families pay 25% more for their food due to acreage taken out of food production by biofuels and other renewable energy sources.

All of the increases in living expenses associated with renewable energy take a disproportionate toll on low-income families. The term 'energy poor' has arisen to describe households forced to spend more than 10% of their income to cover energy costs. Even in the United States, it is estimated that as many as one in four households are now energy poor due to expensive renewable energy policies.[35] One third of Germans and half of the Greeks are now classified as energy poor. In the United Kingdom, environmentalists gleefully point out that higher energy prices have had the desired effects of lowering energy consumption and emissions. However, they fail to point out that 15,000 people in the U.K. died in the winter of 2014-2015 because they couldn't afford to heat their homes.

3) Americans are paying higher taxes. At all levels, solar energy only becomes viable when funded through enormous government subsidies and tax breaks. At the single home level, these subsidies can amount to 40% or $16,000 *per home*. Those who don't install solar (i.e. middle and low income families) are paying billions of dollars per year to those who do (i.e. the well-to-do), providing a huge incentive for putting fossil fuels out of business. The billions doled out to individual homeowners are nothing compared to the tens of billions that the federal government has been throwing at solar energy and other 'green' companies.[36] An electric car company *in Finland* (the now bankrupt Fisker) got $529 million of U.S. tax dollars that will never be repaid. Tesla received a loan of $465 million. The flagship for President Obama's solar subsidies was Solyndra. Solyndra received a government loan of $535 million, went bankrupt, defaulted on the loan, and cost taxpayers the entire $535 million.

4) Businesses and jobs are lost. Although losses are greatest in the energy sector, all segments of the United States economy are adversely affected by higher energy costs. Coal production has dropped by 34% since 2011[37], with West Virginia alone seeing a 45% drop. Since 2012, 332 coalmines have closed, and 27 mining companies have filed for bankruptcy. Over 60,000 coal jobs have been lost. The percentage of working age adults who have jobs in West Virginia has dropped to 53%, which is the lowest in the nation. Needless prohibitions against pipelines and offshore drilling have cost the oil and gas industries hundreds of thousands of jobs.

While progressives like to brag that they don't mind paying twice as much for energy if it saves the planet, individual homeowner only consume 11% of the energy used in America. Businesses and industrial concerns are in a different position. Heavy industries that consume 25% of our energy to produce aluminum, iron and steel, glass, cement, chemicals, paper products, and food largely rely on fossil fuels, much of which is consumed on-site. These industries cannot compete if their energy costs are even doubled, forcing them to either close or move to countries where energy prices and regulatory climates are more cost competitive.

The United States is paying a high price in order to have no impact on global warming.

The Cap-and-Trade Strategy for Eliminating the Use of Fossil Fuels

The only winner in the renewable energy sweepstakes is Big Government. Under the guise of 'saving the planet,' governments are now in a position where they can micromanage entire economies. This power grab is not limited to controlling companies that *produce* energy (see above), but is now being extended to any companies or even individuals who *use* energy. The most visible tool used to assert widespread economic control is so-called 'cap-and-trade' legislation. The shining example of how cap-and-trade is exploited by governments is provided by the state of California.[38,39]

Steps in the cap-and-trade strategy include: 1) Set arbitrary and unrealistic goals for reducing the use of fossil fuels. In California, the goal is to drop CO_2 emissions by a factor of 5 (an 80% reduction) by the year 2050 (only 32 years from now) in order to 'halt climate change.' By 2030 (12 years from now), the goal is to drop emissions by 40%. As California drivers lead the nation in gasoline consumption, and as most electricity is still provided by fossil fuels, it is difficult to envision how this goal is to be met. 2) Set arbitrary and unrealistic *caps* and permit requirements limiting the amount of CO_2-generating energy that companies and businesses can use. The caps are set to dictate that the unrealistic emissions goals are met. Everyone knows that staying under the imposed caps will be impossible. 3) Define a punishing tax structure (called '*trades*') that will allow companies to exceed the caps provided that they pay ever-escalating fees (i.e. extortion) to the government in order to stay in business. According to California's cap-and-trade legislation, these fees are *supposed* to be used to fund renewable energy projects. The fee structure in California is currently generating an additional \$2.5 B/year of revenue collected from 'evil' energy users plus an increase in gasoline prices by 11 cents per gallon. 4) Establish a government bureaucracy (called the Air Resources Board) with the power to decide which companies are exempt from the cap-and-trade policies and which companies will be forced to comply. In effect, this bureaucracy has the power to pick the winners and losers in California's economy on the basis of 'political correctness.'

What have California's cap-and-trade policies accomplished so far? Results include: 1) Emissions in California are currently dropping at a rate of 0.3% per year[40] largely due to the government-mandated construction of several new solar energy power plants.

If this rate continues, emissions will drop by 3.5% by 2030 or less than one tenth of cap-and-trade goals. Gasoline consumption for transportation, accounting for 40% of state emissions, has actually been increasing. In the context of California, let alone the United States or the World, cap-and-trade policies in California have had no impact on global warming. 2) The funding raised by cap-and-trade taxation has more to do with political pork than emissions reductions. Around 60% of the funds raised are not being spent on renewable energy as promised, but on other government priorities including Governor Brown's pet bullet train project (25%) and 'affordable housing' (20%). 3) Companies that a) cannot obtain scarce carbon permits, or b) cannot afford to pay excessive carbon taxes, or c) cannot afford escalating energy bills are simply moving their businesses out of California to more economically-friendly locales.

The exodus of energy-intensive industries has consequences. Politicians can crow that the loss of energy intensive industries has lowered emissions in California. However, these emissions, along with the businesses that create them, have just been transferred elsewhere. Ironically, California's policies are actually leading to an increase in worldwide emissions, as some companies are moving to places having lower environmental standards than California such as China, whose emissions per unit output are twice as high.

Emissions are not the only items being transferred. California is being forced to come to grips with the fact that their state does not control the economy of the entire United States, let alone the entire planet. The loss of industry is having a negative impact on California's economy, resulting in lost jobs and tax revenue. To stem what has been called 'industrial leakage,' California's Air Resources Board has been forced to hand out 'leakage assistance passes' that exempt selected companies from having to comply with cap-and-trade rules. Of course, if companies are not forced to comply, emissions goals cannot possibly be met.

To gloss over this fatal flaw, the government is claiming that all passes will be disallowed after 2030. Are the prospects of a short-term deferment of higher taxes, energy costs, and regulations enticing companies to flock to California? Obviously not! Therefore, California is exploring means for 'leveling the playing field' by forcing the rest of the world to comply with their policies.[41] State legislators have proposed mandating a border-adjusted carbon tax. This means that all companies located *outside* of California would be forced to pay the same cap-and-trade taxes paid by companies located *inside* California. They claim that this would eliminate any incentives for leaving the state. Just imagine how much new state revenue these taxes would generate. The fact that such a carbon tax would violate countless trade agreements as well as the U.S. Constitution seems to be beside the point. Legislators also seem to be oblivious to the economic reality that the massive loss in trade stimulated by their taxes would more than compensate for any new tax revenues collected.

Cap-and-trade policies are only serving to cap and trade away California's economic future.

Actual Prospects for a Renewable Energy Future

The only practical reason for developing renewable energy is not to try to 'save the planet' from global warming, but preparing for a future in which all fossil fuels have been consumed. Fossil fuels are not renewable. Even though new reserves of oil and coal are constantly being found, it is estimated that existing fossil fuels could be exhausted in as little as 200 years (see Chapter 3). Then what? The United States will somehow have to replace the resources that currently provide 82% of our current energy portfolio.

Can renewable energy technologies fill the yawning gap that will remain when fossil fuel reserves are consumed? Unfortunately, an evaluation of existing technologies suggests that the answer is no.

As discussed above, biofuels consume more energy than they produce and are of no help in filling the gap. Hydroelectric power is renewable. However, most of the productive sites for dams are already occupied. Even if the attacks from environmentalists against hydroelectric facilities cease, it is unlikely that hydroelectric power will ever represent more than 3% of our energy needs. That leaves the renewable technologies of wind and solar power.

Even if one ignores the high costs of solar- and wind-generated electricity, as well as technical difficulties associated with the transmission and storage of additional solar and wind capacity, estimates can be made regarding the extent to which renewable energy can replace fossil fuels. Elimination of fossil fuels will dictate that all energy in the future will have to be based on electricity. At this point in time, much of our electricity is generated within 7,700 power plants each having at least 1,000 MW of usable capacity. Moving *all* energy (Fig. 8.1) onto the electrical power grid will require the equivalent of at least 20,000 of such power plants. How much of this capacity can wind and solar energy potentially provide?

As discussed above, the average area required to produce 1,000 MW of actual capacity from a wind farm is 880 square miles. If one makes the unrealistic assumption that wind farms operate at full capacity 100% of the time, the land requirement drops to 350 mi². Using wind farms to provide America's current energy needs of 97.5 quads will require a total land area of at least 11 million square miles. The bad news is that the total land area of the continental United States is only 3 million square miles. The good news is that some of the land between wind turbines can be used for crops, allowing for a partial recovery of some of the total wind farm acreage. For the sake of argument, assume that the entire wind-rich region of the United States from Texas to North Dakota, accounting for 30% of the continental area, was allocated to wind farms. Even here, the maximum amount energy that wind farms could produce would be 8% of the current U.S. energy total, or 5 times what wind farms are producing today.

Solar power consumes 7 times less land than wind power. This means that solar power would 'only' consume 1.9 million square miles within the United States in order to match our current rate of energy consumption. As with wind, some of this land can be put to dual use, such as solar installations placed on the roofs of homes. To be extreme, assume that the entire 106,000 square miles encompassed by all urban areas[42] could be covered with solar panels. Such coverage would account for 5% of our energy needs, or ten times what solar energy is currently providing. If all land in the Southwestern states of Arizona and New Mexico was completely allocated to large stand-alone solar power plants, another 240,000 square miles could be added, providing a grand total of 346,000 square miles or 12% of the continental United States. Even in this totally unrealistic scenario, solar power would provide only 18% of the energy that the United States needs. Coming back down to reality, it is hard to envision that the sum of the absolute maximum totals for solar and wind power will ever contribute more than 20% of our current energy needs. Where is the remaining 80% going to come from? Will the current era of abundant and affordable energy soon become a thing of the past?

The Progressive Solution for a Renewable Energy Future: The 'Sustainable' Economy

The socialist movement paints a picture of a utopian future in which a benevolent and infallible government oversees all aspects of energy and the economy. This future is often referred to as 'the new normal.' In their desired society of the future, all people will be truly equal, as the greedy rich will no longer be able to exploit the masses. All businesses will produce goods at cost rather than for a profit, saving everyone untold resources. People will only use the energy that they really need. Businesses will only produce goods that people really need. The government will be the instrument that decides what people and businesses really need.

This utopian solution is called communism. However, a more palatable definition that progressives use to win over the public is the *sustainable economy*. In a sustainable economy, there is no need for any energy other than renewable energy. Because all of the energy will be renewable, this economy will no longer rely on energy sources that harm the environment such as the fossil fuels that cause global warming. All energy will be 'green', and the planet will be a safer place that will allow all humans, plants, and animals to live in perfect harmony.

The sober reality is that people behind the climate change movement have no intention of 'renewing' any of the energy resources that they destroy. They detest affordable energy, as it facilitates capitalism, free markets, and democracy rather than a massive socialistic or communist dictatorship. Their mission is to turn out the lights all across America. Unless someone stops them, this goal will be accomplished within the next generation.

As outlined above, there is no way that renewable energy sources will ever provide more than 20% or one fifth of the energy that the United States currently uses. If all energy is to be renewable, this means that their plan is for the 'benevolent' government to force the country to slash energy consumption by at least 80%. As industry currently uses almost 90% of our energy, this means that industrial production must also be cut by at least 80%. As industry supplies most of the jobs that provide for the economic well-being of America, gainful employment must also be slashed. The net result is that life for most Americans will be equal — equally bad — with economic and energy resources that are comparable to those in Third-World African states.

Implementation of the Sustainable Economy: Government Coercion

Are you willing to give up 80% (or retain only one fifth) of the energy that you now use to: heat and cool your home, preserve and cook your food, wash and clean your clothes, and drive your car in order to achieve a 'green' economy? If you own a steel mill, can you stay in business if you only operate your blast furnaces for one day a week? Are you willing to have the government decide who gets access to energy and who doesn't? Progressives know that the answer to all of these questions is an emphatic: No! Therefore, they have decided that people should no longer have the freedom to choose their energy future. They believe that it is time for the government to take total control over energy.

Government-enforced transitions to a 'sustainable' economy have already started. The first step in the process is to eliminate the use of fossil fuels and all non-government sources of energy. The elimination of fossil fuels is progressing in stages in the hopes that no one will notice. One place where an aggressive shift to solar and wind power is being vigorously pursued is California. The state is forcing public utilities to start closing fossil fuel plants and to invest heavily in new solar and wind facilities. To date, this government coercion has resulted[40] in: 1) a reduction in state fossil fuel emissions by a whopping 0.3% per year, 2) enormous rate hikes in utility bills, exceeding 25 cents per kilowatt hour for some and with industrial customers facing a 50% increase in utility bills in 2018 alone, and 3) ever increasing rolling blackouts that the state is using to back demands for curtailing energy use.

The state has now started to eliminate the freedom of choice for energy use by individual customers. For example, the Energy Commission in California recently mandated that solar panels must be installed on all new homes[41] (at a cost of up to $12,000) and must be cleaned, inspected, and maintained each year (costing over $1,000/year) to meet state requirements. No doubt the next step will be to demand that all existing homes and businesses retrofit their properties with solar panels, which will even be more expensive. Note that even these measures will have a minimal impact on the CO_2 emissions of California, let alone the world.

The only way for California to really reduce its emissions is to target the use of automobiles. It won't be long before California outlaws the internal combustion engine. Once all people are forced to use electric cars, then all energy used for transportation will be transferred to the electric power grid. Progressive politicians do not seem to realize and/or care that the existing power grid is struggling to meet existing demands, let alone the doubling of demand that would be imposed by the switch to electric cars. In fact, these same politicians can hardly wait to outlaw cars altogether, by making the valid (by then) claim that neither the electricity generation nor grid capacity will be able to come close to meeting demands for electricity-driven transportation.

Once the government controls all energy, it is a simple matter for the government to cut energy usage until the goals of the 'sustainable economy' are met. To help with this mission, progressive politicians have already allocated tens of billions of dollars to develop what they are calling a 'smart' electrical grid. The smart grid is not designed to provide more electrical capacity, but to provide the means for monitoring, policing, and restricting the use of all energy. If implemented, the 'smart grid' will not only control how much electricity your home or business can receive, but will allow the government to control how much electricity goes to each individual appliance. In fact, the government will then have the power to withhold energy from any individuals who are not deemed to be 'politically correct.' You will no longer be in control of your own thermostat. *Big Brother will.*

Future Prospects for Nuclear Power

There is only one viable energy source left in our current energy portfolio that can replace fossil fuels as an alternative to the grim 'sustainable' economy described above. This source is nuclear power, which currently supplies 9% of the energy needs of the United States. Nuclear fuel represents the most concentrated form of energy on Earth. All existing nuclear reactors rely on fission reactions that harvest the energy associated with the splitting of atoms such as uranium (U) to form lighter elements such as strontium (Sr) and xenon (Xe) having lower atomic weights via reactions such as[43]:

$$_{235}U^{92} + {}_0n^1 \text{ (a neutron)} \rightarrow {}_{38}Sr^{70} + {}_{54}Xe^{143} + 3 \, {}_0n^1, \text{ E} = 200 \text{ MeV}$$
(8.1)

The above equation shows that the splitting of a single uranium atom releases an incredible 200 million electron volts (MeV) of energy. This is why atomic weapons are so powerful.

To put fission energy into perspective, when an atom in a solar energy cell harvests a single photon (light ray), the amount of energy captured is around 2 eV, or 100 million times less energy than the energy associated with a single fission event. This explains why ten metric tons of fuel, having a volume of only 18 cubic feet (a cube that is 2.6 feet on a side), provides 400,000 million kilowatt-hours of energy. The land area occupied by nuclear plants is negligible relative to that needed for solar and wind power. As another advantage, nuclear power produces no emissions or greenhouse gases.

Unfortunately, nuclear power — representing our second most important energy source after fossil fuels — is equally vilified by the progressive movement. Environmentalist attacks on nuclear reactors in general and breeder reactors in particular have been so relentless and virulent that neither reactor type is currently under construction in the United States. In fact, due to crippling government regulations, no new nuclear power reactors have been built since 1978. The progressive movement, amplified by the media, has been remarkably successful in scaring the public into accepting that: 1) the world is threatened by destruction and death due to nuclear accidents, and 2) the Earth's environments are being threatened with dangerous contamination due to nuclear wastes.

In Defense of Nuclear Power

The truth is that nuclear power represents an exceptionally safe means of producing energy. Consider the three most notorious reactor failures in world history.[44] Not a single person died as a result of America's worst reactor accident at Three Mile Island. A magnitude 9 earthquake and a subsequent 50-foot high tsunami were required to disable Japan's Fukushima reactor complex on March 11, 2011. This natural disaster killed 20,000 people. However, even though three reactor cores melted down, there was not a single fatality on the reactor site. Even within the plant, only six workers were exposed to radiation levels above those allowed for radiation workers, and not a single person received a dose sufficient to cause radiation sickness. The worst reactor accident in history on April 26, 1986 at the Chernobyl facility in the Ukraine resulted in the deaths of 31 individuals. These people were all plant personnel or rescue workers working at the immediate reactor site, with 29 dying of radiation poisoning. To put this half a century record into perspective, all of these reactor incidents combined have yet to kill as many people as the 111 persons killed by falling debris and other incidents at wind farms within a much shorter period of time.[19]

There are multiple reasons why the safe disposal of nuclear wastes is a political rather than a technical problem. First, because nuclear fuel has such a high energy density, the amount of radioactive waste that is generated is actually quite small. The primary waste is spent uranium fuel rods. The entire inventory of these rods generated over the past 50 years weighs 65,000 tons and has a volume of 4,300 cubic yards. This entire volume is equivalent to a cube that is 16 yards on a side. Second, a safe centralized underground repository for high level wastes already exists in the state of Nevada. The Yucca Mountain Waste Repository adjacent to the Nevada Test Site is completely isolated from human habitation. It is 1,000 feet underground, in an extremely arid climate, and has been scientifically certified as a safe place to store wastes for up to one million years. Third, a broad spectrum of stabilized waste forms have been developed that are more resistant to aqueous attack than granite, and would prevent radionuclides from leaving the site even in the unlikely event that it became flooded. Fourth, if breeder reactors are built, almost all spent fuel could be recycled rather than stored as waste. In spite of all of this, Senator Harry Reid and President Obama prohibited the shipment and storage of any nuclear wastes to Yucca Mountain. The bottom line is that environmentalists and politicians have blocked all attempts leading to the safe storage of nuclear wastes. They then have the audacity to claim that nuclear power is not viable because the nuclear wastes cannot be safely stored.

From a practical rather than a political or emotional perspective, nuclear power is the only *existing* option that can replace fossil fuels and provide for the future energy needs of the United States. However, like fossil fuels, it is important to point out that uranium is a non-renewable resource. At first glance, the availability of uranium in economically ($260/kg U) minable ore of 7.6 million metric tons might appear to be a major technical limitation to an expansion of nuclear power. The Nuclear Energy Agency projects that at current rates of fuel consumption in light water reactors, the world's uranium supplies could be consumed in as little as 230 years.[46] With fuel enrichment and spent fuel reprocessing, supply lifetimes could be doubled to 460 years. However, even with these measures, this estimate suggests that the footprint for nuclear energy might be difficult to expand to meet the world's future energy needs.

Fortunately, the future of nuclear energy is not limited to the current combination of uranium mines and light water reactors. Seawater contains a total dissolved inventory of 4.5 billion metric tons of uranium. Although currently more expensive than mining, technologies exist for extracting uranium from seawater.[47] In fact, uranium can already be harvested from seawater at a cost of $660/kg U. While over twice as expensive as extraction from low-grade ore, this price (when translated to kilowatt-hours) is still less than the real cost of solar and wind energy. In existing light water reactors and current use rates, the uranium in seawater would suffice to power existing reactors for 60,000 years.

Second, advanced reactor designs called breeder reactors actually produce more fuel than they consume. Here, non-radioactive isotopes called 'fertile' materials capture neutrons emitted from radioactive or fissile materials to create more radioactive fuel. The fertile material in existing breeder reactors is uranium. However, another fertile material under investigation is thorium, which is converted into uranium by a nuclear reaction within the breeder reactor fuel. Minable thorium reserves are estimated at 6.3 million tons,[48] which would increase minable nuclear fuel reserves by almost 40%. Even with only uranium as the fertile material, it is estimated that breeder reactors could extend the lifetime of the world's mined uranium supplies to 30,000 years at current use rates. The combination of uranium from seawater and the use of breeder reactors would prevent uranium stocks from being depleted for almost two billion years.

Fusion Reactors: The Ultimate Solution to the Earth's Energy Needs

The energy contained in fossil fuels originated from the Sun (i.e. photosynthesis, see Chapter 7). In the future, it is likely that the ultimate 'solar energy' technologies will harness the nuclear fusion reactions that *power* the Sun, such as[49]:

$$_1H^2 + {}_1H^2 \rightarrow {}_1H^3 + {}_1H^1, E = 4.03 \text{ MeV} \quad (8.2)$$
$$_1H^2 + {}_1H^3 \rightarrow {}_2He^4 + {}_0n^1, E = 17.6 \text{ MeV} \quad (8.3)$$

In the above reactions, $_1H^1$ is the most abundant isotope of hydrogen, $_1H^2$ is the non-radioactive deuterium isotope, $_1H^3$ is tritium, and $_2He^4$ is helium gas. The net reaction cycle shown (the sum of reactions 8.2 and 8.3) consumes three hydrogen atoms to form a helium atom, a hydrogen atom, and a neutron. The net reaction releases over 21 MeV of energy, or ten million times the energy per atom captured in light-harvesting solar cells.

While the energy released in fusion reactions is enormous, it is still less than one tenth of the energy released by the fission of uranium in a conventional nuclear reactor. Then what makes fusion energy the ultimate solution to the world's energy needs? The answer has to do with the fuel used in a fusion reactor. *The fuel is water.* Deuterium (D) is present in the form of the protons in water (D_2O in H_2O) at a natural abundance of 0.015% (one part in 6700).[50] The amount of water in the oceans is 1.7×10^{18} Mton (or 1.7 million billion billion tons). This means that the amount of D_2O (deuterated water) in the oceans is 2.6×10^{14} Mton (260 billion billion tons). Given the vast amount of energy per atom and the vast number of atoms available, the amount of energy represented by fusion energy on Earth is truly limitless.

If fusion reactors are so attractive, then why don't they already exist? The answer is that while fusion energy powers the Sun, fusion reactions require conditions of extreme temperatures and pressures similar to those found within the Sun. No known material can withstand these conditions. In spite of this, scientists are coming surprisingly close to creating the world's first fusion reactors. In current designs, high magnetic fields are being used to suspend and compress small quantities of ionized gaseous deuterium fuel to the point where fusion can occur. The materials that comprise the reactor are far enough away from the heat released by the resulting fusion reactions that the thermal energy can be harvested without destroying the reactor. While most research on fusion reactors has been conducted in government laboratories and universities, the 'Skunk Works' at Lockheed Martin Company recently reported that it is pursuing a compact reactor design that could be ready to commercialize in less than ten years.[51] The reactor is small enough to fit on a semi-truck yet should provide enough power to supply a city of 50,000 to 100,000 people. Lockheed Martin projects that their fusion reactors could cut the cost for desalinating water by 60%.

While it is unlikely that commercial fusion reactors will really be available in ten years, it is probable that fusion energy will be widely available before all fossil fuels are consumed. Fusion reactors open up the prospect for limitless affordable energy without the environmental penalties associated with fossil fuels. Humanity can look forward to a future in which electric cars and affordable clean water are available over much of the world.

Unfortunately, for this very reason it won't be long before progressives begin to shift their relentless attacks on fossil fuels and nuclear reactors to attacks on fusion energy. One possible avenue of attack might be to claim that fusion reactors also generate radioactive wastes that are especially dangerous and deadly. The truth is that the only fusion-produced radioactive byproduct is tritium ($_1H^3$ or T, see Eq. 8.3). Tritium is short-lived (with a half life of only 12.3 years) and decays to background levels in a very short time. The quantities of tritium generated per unit energy are miniscule. In addition, tritium is also easy to store, and in a stabilized form poses no threat to humans.

Summary

The environmentalists' dream that renewable energy sources will replace fossil fuels and prevent man-made global warming is just that: a dream. Even one of the most rabid global warming advocates, ex-NASA manager James Hansen, who oversaw the tampering of government temperature records (see Chapter 9), admitted that: "Renewable energies are grossly inadequate for our energy needs now and in the foreseeable future." Even if pushed to the limit, the impact of renewable energy on global carbon emissions will be negligible. The major threat to humanity is not global warming, but the elimination of abundant and affordable energy based on fossil fuels. Nuclear energy represents the only known option for replacing our valuable fossil fuel resources. Reactors based on nuclear fusion have the potential to provide future societies with almost unlimited energy at a minimal environmental penalty.

Chapter 9: The Perversion of Climate Science

Myth: A group of noble and objective scientists has conclusively proven that the temperature of the Earth is spiraling out of control due to the burning of fossil fuels. Over 97% of all scientists now agree that man-made global warming is a fact. The science is settled.

The Scientific Consensus Regarding Global Warming

During the Dark Ages, 97% of all scholars swore under oath that the Earth was flat. The other 3% was executed for heresy.

Communist governments may be able to claim that they have the support of 97% of the people, but such levels of agreement are impossible in a free society. More tellingly, progressives don't seem to understand that science is not a popularity contest, opinion poll, or election. Scientific conclusions are supposed to be based on provable scientific facts. Instead of citing proven results, climate change advocates would rather gleefully point to a survey reported by John Cook (May, 2013) stating that 97% of all scientists are strong proponents of man-made global warming. However, a closer look exposes yet another myth.[1] The survey was 'Cooked.'

In the United States alone, there are over one million physical scientists. Yet the Cook survey was based on a selection of only 12,000 research abstracts published from 1991-2011 that contained the words 'global warming.' The scientists who wrote these abstracts were predominantly progressive environmentalists. No concrete position on global warming was stated in 8,000 of the abstracts. While 4,000 agreed that the planet was probably warming, only a few specifically stated that humans were the cause. Only 3% stated that there was no such thing as man-made global warming. The conclusion that progressives and the media draw from the Cook survey is that since only 3% of the abstracts made a definitive statement against man-made global warming, the other 97% of scientists obviously agree that it represents the truth.

In contrast to the Cook survey, 4,000 scientists, including 72 Nobel Prize winners, signed the Heidelberg appeal clearly stating that there is no scientific basis behind man-made global warming.[2] Over *thirty one thousand* American scientists have signed the Oregon Petition[3] stating that: "Proposed limits to greenhouse gases would harm the environment, hinder the advance of science and technology, and damage the health and welfare of mankind. There is no convincing scientific evidence that human release of carbon dioxide, methane, or other greenhouse gases is causing or will, in the foreseeable future, cause catastrophic heating of the Earth's atmosphere and disruption of the Earth's climate." All of these scientists (100%) believe that global warming represents politics and not science. If science really was decided by elections, polls, petitions, or surveys, the recorded 'votes' cast in clear support of global warming (4,000 in the Cook survey) are to be compared with the 35,000 votes that have been cast in anti-global warming petitions. The grand total of this broader 'survey' indicates that 90% of 'all' scientists agree that progressive claims regarding man-made global warming are a sham.

Prominent scientists who have spoken out against human-induced global warming include:

- *Professor Roger Revelle:* Professor Revelle was a coauthor on the original paper in 1957 postulating that carbon dioxide from fossil fuel emissions might have an impact on the earth's climate.[4] Al Gore claimed Professor Revelle as his mentor. Professor Revelle is widely regarded as the father of the global warming movement. Near the end of his life in 1991, he began to have serious reservations regarding his own theory, expressing those doubts in an article published in *Cosmos Magazine,*[5,6] In 1988, even before Earthly temperatures had ceased to rise, he wrote: "My own personal belief is that we should wait another 10 or 20 years to really be convinced that the greenhouse effect is going to be important for human beings in both positive and negative ways." Had he lived to see temperatures level off or even decrease, we wonder what he would have said.

- *John Coleman:* John Coleman had a fifty year career as a meteorologist and television weather anchor on shows such as ABC's *Good Morning America*. His is the founder of the Weather Channel. He said,[7] "If you get down to hard, cold facts, there's no question about it: Climate change is not happening. There is no significant man-made global warming now, there hasn't been in the past, and there's no reason to expect any in the future."

- *Professor Ivar Giaever* (University of Oslo): Professor Giaever won the 1973 Nobel Prize in Physics. Professor Giaever labels the global warming movement as pseudoscience,[8] explaining that: "In pseudoscience, you begin with a hypothesis which is very appealing to you, and then you only look for things which confirm the hypothesis." He disputes that global warming is "incontrovertible." "See, that's a religion. That's a religious statement like the Catholic Church says the world is not round."

- *Professor Harold Lewis,* Emeritus Professor of Physics at the University of California, Santa Barbara[9]: "The global warming scam, with the (literally) trillions of dollars driving it, has corrupted so many scientists. It is the greatest and most successful pseudoscientific fraud I have seen in my long life as a physicist."

- *Professor Jonathan Jones,* Professor of Physics at Oxford University[10]: "My whole involvement has always been driven by concerns about the corruption of science. I was looking up some minor detail about the Medieval Warm Period and discovered this weird parallel universe of people who apparently didn't believe it had happened, and even more bizarrely appeared to believe that essentially nothing had happened in the world before the 20th century. The first extraordinary thing about the evidence [for global warming] was how extraordinarily weak it was, and the second extraordinary thing was how desperate its defenders were to hide this fact."

- *Professor Richard Lindzen,* Alfred P. Sloan Professor of Meteorology at the Massachusetts Institute of Technology on the primary evidence for global warming[11] (see Climategate below): "The documents themselves are not speculation. They are unambiguously dealing with things that are unethical and in many cases illegal. There is no point in any scientific group endorsing this. We are not crooks. And yet if we endorse this we are becoming that."

- Dr. John Christie: Nobel Prize winner Christie conducts research in the Earth System Science Center at the University of Alabama. He heads the program responsible for collecting satellite data regarding the earth's temperature. His statements include[12]: "I am still a strong critic of scientists who make catastrophic predictions of huge increases in global temperatures and tremendous rises in sea levels." "I see neither the developing catastrophe nor the smoking gun proving that human activity is to blame for the warming we see … From my analysis, the actions being considered to "stop global warming" will have an imperceptible impact on whatever the climate will do, while making energy more expensive, and thus have a negative impact on the economy on the whole." Dr. Christie has backed up his statements with hard data showing that the earth's temperature is not rising nearly as rapidly as the IPCC global warming computer programs have predicted (Fig. 9.1), and debunking numerous claims made by the media regarding how global warming is generating extreme weather (see Chapter 6).

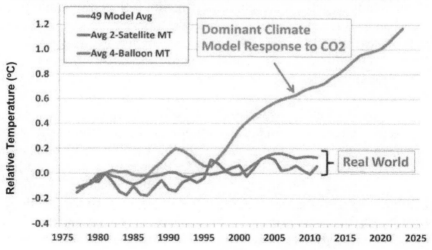

Fig. 9.1 Graphs of the relative temperature change (in °C) predicted by climate computer models compared with temperature changes that have actually been observed since 1975 (adapted from data presented in Ref. 13).

None of these scientists believe that the science behind global warming has been 'settled.'

The Perversion of Climate Science: Climategate

Scientists in the 'climate denier' camp are upset with the global warming movement for multiple reasons. Global warming proponents have totally abandoned the scientific method.

Science has degenerated into an exercise in socialist propaganda. The case of Professor Michael Mann provides a perfect example of how climate change advocates have been handling the 'science' of global warming in the modern era.

Peer reviewed studies by over 750 scientists from over 450 research institutions in 40 countries were used by Dr. Hubert Lamb, the founder of the Climatic Research Unit at East Anglia in the United Kingdom (CRU), to compile a complete set of climate data spanning the past 1000 years[14,15] (Fig. 9.2). This vast data set, used by the International Panel on Climate Change (IPCC) as recently as 1990,[16] clearly shows the climatic periods known as the Medieval Warm Period and the Little Ice Age. Miraculously, the climate change movement has succeeded in having all of this data swept aside and replaced by the now famous 'hockey stick' curve (Fig. 9.2) representing the work of a single researcher: Professor Michael Mann.[17] Did this dramatic transformation represent an example of exceptional science? Unfortunately, this question has multiple answers, all of which are appalling.

Professor John Christie testified before Congress[18] that Mann "misrepresented the temperature record of the past thousand years by: a) promoting his own result as the best estimate, b) neglecting studies that contradicted his, and c) amputating another's result so as to eliminate conflicting data." Unfortunately, Professor Christie's comments do not come close to capturing the depth and breadth of fraud and corruption that surrounds Mann and the entire global warming movement. The actual means by which Professor Mann's 'hockey stick' graph became the central image of the movement were clearly revealed in November of 2009[19] when thousands of emails, documents, and data source code were leaked to the public from within the Climatic Research Unit under director Phil Jones. The releases involved both CRU and the major climate institution in the United States headed by Professor Mann at Penn State University. These releases, as well as the subsequent trials for scientific fraud against both groups and NASA, form the basis for Climategate scandal revelations regarding the tactics of the global warming movement.

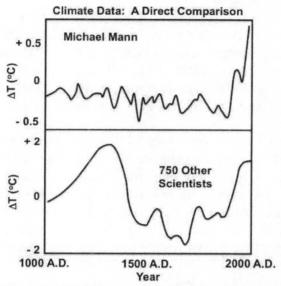

Fig. 9.2 A direct comparison between Professor Michael Mann' famous 'hockey stick' climate curve (top, adapted from data in Ref. 20) and a curve representing a compilation of data from over 750 other climate scientists that was used by the IPCC as recently as 1990 (bottom, adapted from data in Ref. 21).

All the following "inconvenient truths" stem from the leaked emails and records:

1) Falsification of Climate Results

In public, CRU and Penn State scientists are sticking to their predictions that the present decade is the warmest ever, and that the science of global warming has been 'settled.' However, their private emails reveal that these climate scientists clearly recognized that temperatures had not risen for the previous 15 years and in fact had been falling for nine years. CRU head Phil Jones was forced to admit *under oath* that the climate data show no warming since

1995, and conceded that there is no proof that any warming period has been caused by human activities.

Even when faced with the knowledge that their model predictions were incorrect, scientists at CRU and Penn State continued to try to rig climate results in favor of global warming using strategies such as[19]:

- Scientists were complicit in the deliberate placement of temperature monitoring stations near man-made heat sources (e.g. air conditioning ducts) and hot objects (e.g. asphalt pavement in the Arizona desert). Temperature readings from these stations were then given precedence in the data sets used to illustrate rising temperatures.
- Data was thrown out from 75% of the world's temperature reporting stations, with a clear bias toward removal of data from high latitude, high altitude, and rural locations that report colder temperatures. Only 25 of the 600 stations in Canada were used, and the frozen tundra of Siberia went largely unreported.
- The groups admitted to tampering with their own data to eliminate any of their own results that didn't agree with the global warming hypothesis.

Even after this extensive 'editing,' the climate institutes were *still* not getting the temperature increases they desired. To generate his famous 'hockey stick' curve, Professor Mann admitted in an email that he used the trick of 'adding temperatures to each series (of data) to hide the (temperature) decline.' In other words:

Professor Mann falsified even his own data.
The 'hockey stick' represents a work of fiction rather than fact.

2) Cover-Ups of Scientific Fraud

CRU and Penn State were determined to ensure that no other scientists were able to dispute their false claims. They repeatedly denied other scientists any access to their data. Finally, outside scientists were forced to demand the climate data be turned over for independent evaluations under the Freedom of Information Act. In response:

- Contributing scientists were ordered by CRU and Penn State to take the illegal step of destroying their email records, as well as computer codes and relevant data.

- CRU refused to turn over any records, stating that most of the data had been 'thrown out' and that they 'didn't do a thorough job of keeping records.'

- Key scientists also refused to turn over records because 'the data is restricted for academic purposes' and that they were not about to release data that would allow other scientists to examine or pick on their long-standing research.

Hmm. Is that how free and open science is supposed to work?

3) Perversion of the Peer Review Process

The emails show that they took steps to pervert the peer review process that controls what papers get published and what proposals get funded, deliberately black balling those scientists that they knew didn't agree with them. Examples include:

- Emails show that they repeatedly leaned on journal editors to ensure that close colleagues and friends were always selected to review their papers rather than independent scientists.

- Conversely, editors were leaned on to reject any papers that did not express agreement with the CRU/Penn State global warming agenda.

- The CRU/Penn State team accumulated sufficient power that they were able to define the peer review process for leading climate panels such as the United Nations' International Panel for Climate Change (IPCC). Their process was then used to exclude contradictory scientific results from the IPCC's four Assessment Reports.

4) Intimidation of 'Climate Deniers'

When all else fails, climate change advocates rely on intimidation to get their way. Examples from the Climategate files include:

- CRU/Penn State pursued an intense lobbying campaign to have a journal editor removed from his job because he failed to toe the line regarding their global warming agenda.

- The institutes established an expensive website having the specific goal of discrediting and denigrating their scientific opponents. One objective was to smear the reputations of competitors to keep them from receiving further federal funding.

- The climate institutes exchanged emails regarding how they might be able to use their doctored results to 'shake down the oil companies for money' to further their cause even though they were already being funded to the tune of $20 million.

- Michael Mann has filed numerous lawsuits against anyone who dares to challenge him.[22] For example, in 2012 he sued radio talk-show host Mark Steyn and others for "defamation of a Nobel Prize recipient" even though Mann has never been awarded the Nobel Prize.

How did the media handle the Climategate scandal? Did they apply the same investigative zeal that they applied in the anti-Republican Watergate scandal? You know the answer. When the scandal broke, ABC, CBS, and NBC ran a total of zero stories. Since then, media efforts have been applied to covering up the scandal and even rewriting climate history. Common websites such as Wikipedia have been censored to try to discredit the leakers rather than emphasizing the overwhelming evidence of climate fraud. Web sites are commonly being censored and reorganized to downplay references to the up to 100 climate cycles that have occurred in human history, including the Medieval Warm Period and the Little Ice Age.

Falsification of Climate Data

Manipulation and falsification by the Global Warming movement has been applied to any data sets having any links to climate change. Specific examples include the doctoring of temperature records (see below), sea level records (Chapter 5 and below), polar bear populations (Chapter 7), and claims regarding renewable energy (Chapter 8). Regardless of the data set, the overriding central principle is that all actions are to be based on the unshakable (and unverified!) hypothesis that fossil fuel emissions are destroying the planet.

The next step in the chain is to construct models having the goal of convincing the public that this global warming hypothesis is correct. The premise is that the general public believes that computers and models are completely objective and infallible even though both can be made to provide any result that the programmer so desires. In fact, the same simple concept is applied to all models, whether the 'results' are used to describe temperatures, the melting of ice (Chapter 5), sea level changes, or even polar bear populations. The central model states that since fossil fuel emissions have been increasing at an exponential rate, all of the above quantities must also increase (or in the case of polar bear populations decrease) at an exponential rate. After the model calculations and curves have been generated, the next step involves either finding or creating data sets that support the model. It is at this stage where the most egregious and pervasive cases of dishonesty and scientific fraud take place.

Government Falsification of Temperature Records

The climate activists in charge of the National Oceanic and Atmospheric Administration (NOAA) were dumbfounded. Their own temperature data as well as satellite results had been showing that the climate had been stable or cooling since 1998. Even worse, the climate was cooling while CO_2 levels continued to rise. The recent climate was not being politically correct. NOAA was coming under extreme pressure from the media and progressives to reconcile this global cooling with their global warming agenda. As recently as 2014, they were still struggling explain how this 'pause' in warming could occur,[23] including volcanoes, El Nino, or that the excess heat from CO_2-emissions was being 'sucked into the deep ocean where no one could see it.' It was time to resort to the strategy that progressives always use in times of crisis: if you don't like the facts, throw them out and make up your own.

Since December of 2015, the web sites of both NOAA and NASA have simply eliminated what they had been calling a 'pause' in global warming. According to these government agencies, the climate has never experienced a pause. Forget about the pause.

There never was a pause. They have taken their data behind a dark curtain for 'editing' and 'reanalysis' . . . and *viola*. Climate data suddenly show that the Earth's temperature has increased by 1°C in just the last few years. This falsification of government climate data has not gone unnoticed. Rep. Lamar Smith (R-Texas), who runs the House of Representatives science and technology committee, has demanded that NOAA produce their data for independent analysis. Knowing full well that their methods and manipulations would crumble under public scrutiny, NOAA has refused to release the subpoenaed documents. Judicial Watch has sued NOAA under the Freedom of Information Act to obtain access to the NOAA climate data.[24,25] NOAA continues to refuse to turn over the documents. Does this sound familiar?

The manipulation of climate data by NOAA and NASA covers more than just the past decade. The U.S. government's published temperature data for the entire span of years from 1880 to 2010 have been tinkered with 16 times over the past three years alone.[26] As a point reference, examine the climate records reported by NASA as a function of time (Fig. 9.3).

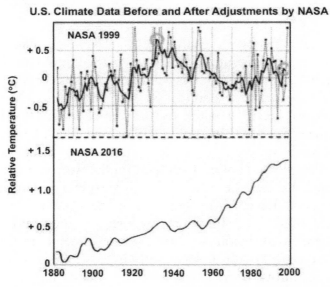

U.S. Climate Data Before and After Adjustments by NASA

Fig. 9.3 A direct comparison of the archival data for average temperatures across the United States as reported by NASA in 1999 and the 2016 curve presumably generated from the same data after government manipulation. From data reported in Ref. 26.

As recently as 1999 (top curve), the major database of temperature records was still more or less intact. The 1999 curve highlights the four most recent major climatic episodes: 1) From 1880 to 1930, temperatures gradually increased by around 1°C (2.3°F). 2) From 1940 until around 1970, temperatures then dropped by around 0.7°C (1.6°F), reducing the temperature almost back to where it was in 1880. 3) Temperatures then rose until the mid-1990s by around 0.5°C. 4) Since 1998, the temperature has been essentially constant.

The progression of actual temperatures does not come close to fitting any of the major hypotheses of the global warming movement. NOAA and NASA were forced to deal with at least three major problems in order to bring climatic data back into 'compliance': 1) Real data show that the world is actually colder today than it was in the 1930s and 1940s (see circles in Fig. 9.3). It is only 0.1°C warmer than it was in 1910. 2) The real climate curve does not look anything like Professor Mann's famous 'hockey stick' curve that forms the basis for the global warming movement. 3) The real climate curve clearly shows that there is no relationship between the Earth's temperature and rising atmospheric CO_2 levels (Fig. 1.1). Clearly something had to be done.

The bottom curve in Fig. 9.3 (NASA 2016) shows what has been 'accomplished' during the past 17 years. The same government agency now reports a progression of climate results that looks completely different even though the curve is presumably based on the same archival temperature records. Amazingly, the climate curve is now in total agreement with the global warming movement. The cooling trend between 1940 and 1970 that led climate alarmists to warn that fossil fuel emissions were plunging the Earth into another Ice Age (see Chapter 1) has been largely eliminated. The data now show that temperatures are exponentially increasing along with rising CO_2 levels in agreement with Professor Mann's 'hockey stick' curve. Not only that, the curve shows that the Earth's temperature has increased by a whopping 1.4°C (~2.5°F) since 1880, showing that the Earth is headed for man-made global annihilation.

How has such a dramatic transformation in the Earth's climate curve been accomplished? As with every activity carried out by the climate change movement, the rule used to generate the new official curve starts with the assumption that the Earth's temperature must scale with atmospheric CO_2 concentrations. Next, government scientists examine temperatures that have actually been reported. Temperatures that are out of line with the prediction are gradually and systematically 'adjusted' and replaced by 'corrected' computer-generated temperatures. In addition, with the exception of North America and Western Europe, accurate temperature records are largely non-existent, so it is easy to create computer-generated temperature records to fill in all blank spaces on the map. The fraction of temperatures that have been 'adjusted' has increased exponentially with time, from around 10% in 1980 to over 50% today.[26-28] The net adjustments that have been made can be visualized by subtracting the actual temperatures from the temperatures NASA is now reporting (Fig. 9.4, top).

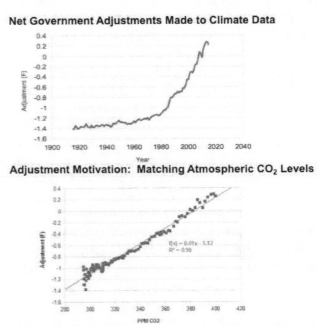

Fig. 9.4 The nature of government tampering is shown.[26] The top curve shows the 'adjustments' that were made to convert the actual archival temperature records of the United States into the curve NASA presented to the public in 2016 (Fig. 9.3). It is no coincidence that the 'adjustments' are an exact match to rising carbon dioxide concentrations (bottom curve).

It is not a coincidence that the 'adjustment' curve resembles the increases that have been observed in atmospheric CO_2

concentrations over the same time period. In fact, a plot of the temperature 'adjustments' versus CO_2 concentrations is a straight line (Fig. 9.4, bottom) *as intended.*

NOAA and NASA are not alone. Tampering has been uncovered in the temperature records of states including New York, Maine, and Alaska. For example, Heat transfer specialist Mike Brakey reports[29]: "I have discovered that between 2013 and 2015, some government bureaucrats have rewritten Maine climate history. This statement is not based on my opinion, but on facts drawn from NOAA 2013 climate data vs. NOAA 2015 climate data after they rewrote it. We need only compare the data. They cooked their own books." It is fortunate that the climate change movement has yet to figure out how to change or eliminate their own archival data except on their own web sites.

Unfortunately, socialist governments eager to support the climate change agenda are doctoring actual temperature records all over the world.[30] On the international scene, changes in temperature records have been uncovered in Australia, Canada, Costa Rica, England, Iceland, New Zealand, and Paraguay. For example, Traust Jonsson, former head of the Iceland climate research office, was stunned to find that the cooling that had damaged his country's economy due to excessive sea ice formation in the 1970s had completely disappeared from the temperature records. If only the media could erase all of the alarming global cooling articles they wrote during that same time period.

A graph providing international evidence of government falsification is provided by the case of New Zealand[31] (Fig. 9.5). New Zealand's National Institute of Water and Atmospheric Research (NIWA) is responsible for keeping the National Climate Database. The bottom curve represents the official government climate report compiled by Dr. Jim Salinger (who worked for CRU in the United Kingdom). The top curve represents the data that independent researcher Richard Treadgold working on behalf of the New Zealand Climate Science Coalition produced using the exact same NIWA database. The temperatures reported in the actual database show that there has been no significant warming in New Zealand since 1850. In contrast, the temperatures reported in the government document suggest that the average temperature in New Zealand has increased by 1.2°C (over 2°F) since 1900. A direct comparison shows that older temperatures were adjusted downward while modern temperatures were adjusted upward to make the government curve. There is nothing in station histories to warrant these adjustments. Repeated requests by multiple scientists to Dr. Salinger as to why such adjustments were made have consistently gone unanswered, mirroring the CRU/Penn State Climategate cover-up.

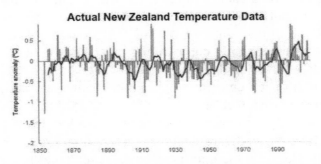

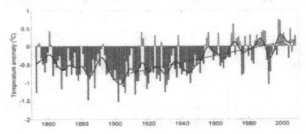

Fig. 9.5 Another example of government tampering is revealed by results that have been reported by government scientists in New Zealand.[31] The top curve shows a compilation of the actual temperature records, while the bottom curve shows the data that have been presented to the public after making the proper 'adjustments.'

Falsification of Sea Level Changes

Sea level records provide wider ranging examples of the methodologies used to promote the global warming agenda.[32] The starting point is to examine complete sets of historical tide gauge reports. As documented in Chapter 5, records that were reported by the IPCC until recently, show that sea levels have been increasing at a slow constant rate of 1.7-1.8 millimeter per year during the

Industrial Age rather than showing the alarming exponential increases predicted by climate models. That will never do.

The global warming movement will stop at nothing to promote their agenda. The bald-faced lie is often the first strategy deployed. Since everyone knows that the 'science of global warming is settled,' any statement consistent with the climate change agenda is expected to be believed without question. Several island nations, including the Maldives, the Solomon Islands, Vanuatu, Tuvalu, and Nauru have succeeded in extorting money from guilty developed nations simply by stating that their islands are about to disappear under the oceans due to global warming.[33] The Australian Government has already sent tens of millions of dollars to the Solomon Islands and Nauru from the 'Pacific Adaptation Strategy Assistance Program' and the 'Green Climate Fund.' Media outlets have taken up the cry with articles such as "Sea Level Rise Has Claimed Five Whole Islands in the Pacific" (Scientific American, May 2016[34]). Unfortunately, neither the media nor Australia bothered to check the validity of these claims. In fact, almost 25 years of tide gauge records gathered by the Australian Bureau of Meteorology[33] show that there has not been any measurable rises in sea levels in any of these locations (Fig. 9.6.).

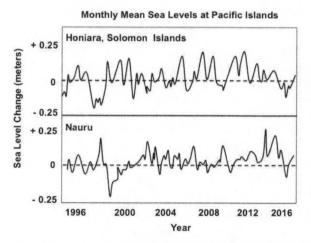

Fig. 9.6 Mean monthly sea levels from 1994 through 2017 at Nauru Island (top) and Honiara in the Solomon Islands (bottom)(adapted from data is Ref. 33). Dashed lines indicate that no change in sea level has occurred.

Afraid that people might want to actually examine sea level records, a fall back strategy involves 'adjusting' tidal records to provide the right answers as has been done with temperature records (Fig. 9.3). As with temperature records, it is fortunate that archival records still exist that plainly show the extent to which sea level data has 'evolved' to provide the proper climate change results. The degree of tampering that has been 'accomplished' with regard to tide gauge data[35,36] since 1983 is shown in Fig. 9.7.

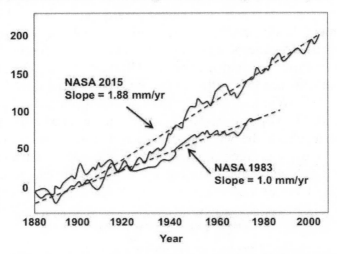

Sea Level Rise from Tide Gauge Results: Complete vs. 'Adjusted'

Fig. 9.7 Tide gauge results as reported most recently by NASA in 2015 (top curve) as compared with a similar report issued by NASA in 1983 (bottom curve)(adapted from data in Ref. 35). Curves similar to the bottom curve were also reported in two other NASA publications in the 1980's. The sea level curve reported by IPCC in 1990 (Fig. 5.8) closely resembles the top curve. Note that both curves are supposedly derived from the same archival tide gauge data.

Notice that prior to tampering, sea level data show that the rate of rise slowed starting around 1950 rather than increasing as predicted. Notice that both before and after tampering, the sea level curve is linear with time rather than showing the exponential increase predicting by the global warming community. Notice that the slope of the line after tampering is almost twice what it was before tampering. Finally, notice that even after tampering, the slope of the sea level rise curve is not steep enough to be alarming (see Chapter 5).

NASA and NOAA have no shame when it comes to doctoring sea level data.[34,36] NOAA has 240 tide gauges. Of these, 86% (6 out of 7) show less sea level rise than the current NOAA claim (based on satellite data, see below) of 3.2-3.4 mm/yr. The average for all NOAA tide gauges is 1.4 mm/yr, or around one third of the reported value. The new 'net sea level rise' is accomplished by eliminating data from gauges that are not providing 'politically correct' answers while giving precedence to those that support the global warming agenda. Is the sea level at a given site actually rising relative to the land, *or is the land on which the gauge is placed falling relative to the level of the sea?* Only 10% of NOAA tide gauges show a sea level rise exceeding their 'average' value of 3.4 mm/yr, and all of those are in places where the land is known to be sinking.[37] The IPCC chose to report results for only one of the six tide gauges in Hong Kong[32] because it showed the largest apparent sea level rise of 2.3 mm/yr. However, geologists have clearly shown that the selected gauge is in an area where the sediment is known to be subsiding. In fact, this particular gauge is the only one that *shouldn't* be used to calculate sea level changes.

If deception doesn't work, then destruction of evidence is deemed to be perfectly justifiable by the climate change community. For example, the documentary *Doomsday Called Off* contained images of a tree growing in the flood plain in the Maldives.[38] The presence of this tree clearly indicated that sea levels had not increased to an appreciable extent since at least 1950.[32] After viewing the documentary, an Australian climate team flew to the Maldives. These so-called environmentalists were then observed ripping the poor tree out of the ground by the roots and then taking pictures at the site.[38-41] Tree? What tree? If there was a tree, it must have 'washed away,' providing further 'proof' that sea levels in the Maldives are rising at an alarming rate.

Perhaps sensing that they have pushed the doctoring of tide gauge results as far as possible, the climate change community has moved on. Since 1993, they have been reporting sea level rises based on satellite altimetry rather than tide gauges. At first glance, this makes perfect sense, as satellites can scan the entire ocean rather than being limited to a few tide gauge locations. Besides, the public trusts satellites. Unfortunately, this trust is grossly misplaced, as global warming advocates are the very people in charge of handling much of the satellite altimetry data (i.e. the fox is guarding the chickens).

The ease with which satellite data can be manipulated is illustrated by satellite results reported by NASA[35] as of 2017 (Fig. 9.8) and by the European Space Agency[42] in late 2011 (Fig. 9.9). Given the precision and global scope of satellite measurements, one might expect to see perfect agreement between the NASA and ESA results. Such agreement is not seen. More alarming is the fact that neither NASA nor the ESA can provide consistent answers from analyses of their own data. *Most* alarming is the fact that both agencies have been able to manipulate their data — which presumably cannot be changed once collected — to inflate their claims for recent sea level increases by a factor of two to three to comply with the politically correct climate narrative. However, each agency has deployed a different strategy to achieve the desired results.

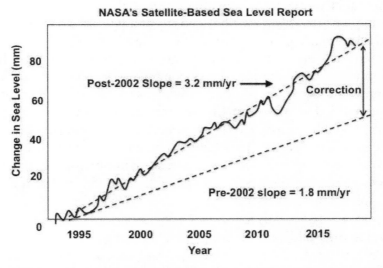

Fig. 9.8 A graph summarizing NASA's report of rising sea levels as of 2017 (adapted from Ref. 35). The dashed line indicates the slope of the curve. The free dashed line indicates the slope reported prior to 2002. The difference between the two dashed lines represents NASA's new 'correction factor.'

Satellite data reported by NASA has always shown that sea levels have been increasing at a constant linear rate (rather than the exponential increase predicted from fossil fuel emissions (Fig.

1.1)). From 1992 to 2002, satellite and tide gauge results reported by NASA and IPCC were in agreement with both indicating a rate of rise of around 1.8 mm/yr. However, starting in 2003, satellite data as reported by NASA and the IPCC suddenly showed a sharp uplift of to produce curves claiming to show that sea levels are now rising by 3.2-3.4 mm/yr. In other words, both groups have now produced new curves by simply adding a line having a slope of from anywhere between 1.5 and 2.3 mm/yr underneath their actual satellite data. The IPCC explained that this uplift was produced using a 'correction factor.' When asked to justify this correction factor by sea level expert Dr. Nils-Axel Morner, the IPCC representative answered truthfully that, "We had to do it. Otherwise we would not have gotten any trend."[32]

Although the rationale for doing so is unclear, the ESA 'corrected' their data to simultaneously solve two problems.

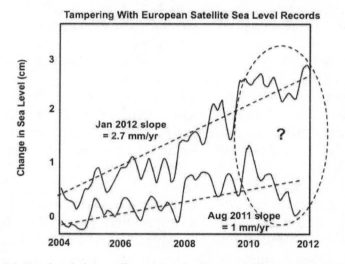

Fig. 9.9 Sea level data collected by the European Space Agency's Envisat satellite as reported on August 4, 2011 (bottom curve) and four months later (top curve)(adapted from Ref. 42). Presumably, both curves were generated from the same data. The reported curves are fairly similar between 2004 and 2009. From 2010 on, the two curves look completely different (see region in dashed oval).

The problem with the overall curve was that sea levels were not rising fast enough to be alarming or to be in line with IPCC model predictions. However, an even more damning observation was that sea levels actually started dropping starting in 2010. Although the climate change movement is loath to admit it, there have been multiple periodic drops in sea levels throughout the modern era. Tide gauges across the eastern seaboard of the United States have registered drops in sea levels by 1 mm/yr since 2009.[43] Even NASA satellites show three drops since 2010, with the first drop coinciding with that shown in the Envisat data. Even according to NASA satellite data, sea levels have been dropping at a rate of more than 1.2 mm/yr since 2015.[44] However, since dropping sea levels are not to be tolerated, the Europeans have simply decided that their data must be 'wrong' and they have accordingly eliminated the drop completely.[42] Note that by eliminating the drop, they have also instantly succeeded in raising the rate of sea level rise since 2004 from 1 mm/yr to 2.7 mm/yr.

When challenged to defend their 'corrections' to satellite data, climate change advocates come up with glib explanations that are too numerous to count in order to win over a gullible public. The most popular explanation used by NASA and the IPCC to defend how the average slope of the satellite data was 'corrected' from as low as 1 mm/yr up to as high as 3.4 mm/y is referred to as a 'Global Isostatic Adjustment.' According to this adjustment, the seabed of the entire ocean has been subsiding or sinking at a constant rate of 1.5-2.5 mm/yr (which is farfetched). To report the 'true' rate of sea level rise, NASA and the IPCC say that they have 'recalibrated' their satellite data by adding the rate of subsistence to the sea level rise that satellites actually observe.[45] Of course, if the seabed is falling as rapidly as claimed, then *the sea level change actually experienced by humans drops back to the slope of 1 mm/yr as indicated in the satellite data prior to the correction.*

Most amusing are the attempts that the global warming community has made to try to explain away dropping sea levels. Just as global warming is said to be the sole cause for rising sea levels since 1880, *climate change advocates now have multiple explanations as to why global warming is now causing sea levels to fall.* For example, based on the unsubstantiated claim that global warming has caused worldwide droughts (see Chapter 6), NASA scientist J. T. Reager developed the following scenario.[46] Worldwide droughts have increased demands placed on water supplies, so since 2009, humans have started to store more water on land for increased 'irrigation and consumption.' These human activities coupled with drought conditions have dried out all soils to cause "the land to act like a sponge." Now when rain falls on land, it no longer flows back into the oceans. Thus the land now removes so much water from the oceans that sea levels are no longer rising as predicted, and may even fall. Hmm. Perhaps if Reager and his colleagues wish to retain any credibility, they should just lay low until sea levels start to rise once again.

The Criminalization of Science and Free Speech

Unfortunately, the progressive movement is now deploying a more ominous approach to 'settle the science.' If they succeed, over 97% of scientists had better say that they believe in man-made global warming if they wish to receive any more federal funding, publish any more papers, or even stay out of jail. The climate change community cannot stand to have their positions challenged. Their ultimate solution is to have the government prosecute anyone who dares to defy them by creating laws that violate the First Amendment of the Constitution. If the science is indeed 'settled,' why should any of this be necessary? What are they afraid of?

Rhode Island senator Sheldon Whitehouse got the ball rolling by calling for the use of RICO (the Racketeer Influenced and Corrupt Organizations statute created to combat the Mafia) to press criminal charges against anyone who speaks out against global warming.[47] Bernie Sanders, Hillary Clinton, and 50 assorted environmental and civil rights groups demanded that Attorney General Loretta Lynch go after Exxon-Mobil for publishing papers in peer reviewed journals that are 'misleading the public' because they do not support global warming. Their 'crime' according to Eric Pooley of the Environmental Defense Fund[48] was that "we've had 20 years of delay because of doubt and confusion sowed by Exxon-Mobil and climate deniers." New York Attorney General and progressive activist Eric Schneiderman beat Loretta Lynch to the punch by filing a major lawsuit against Exxon-Mobil.

As of June, 2018, the cities of San Francisco, Oakland, New York, and Seattle are suing oil giants including British Petroleum, Chevron, Conoco-Phillips, Exxon-Mobile, and Royal Dutch Shell for billions of dollars for the 'crime' of producing, promoting, and selling fossil fuels.[49] The premise is that all of these companies 'must pay' for knowingly contributing to the future destruction of the planet due to global warming. However, if these cities are so concerned about global warming, why don't they outlaw the use of gasoline and other fossil fuels within their own jurisdictions? Why don't they outlaw the use of gasoline-powered automobiles and trucks? Aren't these cities just as 'guilty' as the oil companies for allowing the use of fossil fuels to continue? Shouldn't they be listed as co-defendants in their own lawsuits?

Of course, the liberal politicians in these cities don't dare prohibit the use of fossil fuels. Imagine what would happen in large urban areas such as San Francisco, New York, or Seattle if motor vehicles and fossil fuels were suddenly prohibited. Their economies and standard of living would instantly collapse. People would flee these cities in droves. No one in their right mind would vote to retain any politician responsible for implementing such policies. This is because fossil fuels represent an enormous boon to humanity rather than the bane that is constantly painted by the climate change community.

Fortunately, not a single court has yet to recognize the validity of any claims against the oil companies due to any so-called destruction caused by climate change. First, judges have yet to see a single concrete example of anything that has been 'destroyed.' Second, judges can point to the fact that the federal government and most state governments are encouraging the oil companies to produce fossil fuels for the enormous economic and societal benefits that they provide. Third, judges recognize blatant extortion when they see it. Judges recognize that politicians are trying to shake down the oil companies to prop up declining local economies without being forced to raise taxes. By collecting billions of dollars in damages from oil companies, politicians are also trying to impress their constituents about how environmentally conscious they are. However, many of these same constituents don't seem to realize that these lawsuits will not really have the effect of punishing the 'evil' oil companies. Oil companies will simply recoup their losses by raising gasoline prices. The net effect will be the same as if *everyone* were to suffer yet another increase in gasoline taxes, with poor and middle class people paying the greatest penalty for the irresponsible behavior of a few rogue cities.

Criminal charges have not been limited to large corporations, but include attacks on individual citizens.[50] For example, Professor Michael Mann is pressing criminal charges against conservative radio talk show host Mark Steyn for daring to challenge his data and conclusions.[51] All of these actions violate the First Amendment. All of these actions are frightening. These actions indicate that progressives are willing to do anything, including throwing people in jail, to prevent people from participating in free and open debates on scientific issues that fly in the face of political correctness. This government suppression of scientific ideas resembles how science was conducted in the Soviet Union under Joseph Stalin.

Summary

A group of scientists, government agencies, and media outlets has been doing everything in their power to distort, misrepresent, and falsify climate records in order to promote an agenda that is purely political rather than remotely scientific. The most recent climate curves produced by these individuals 'prove' that the earth is now experiencing a dramatic period of global warming. Unfortunately, this 'warming' has nothing to do with fossil fuel emissions and everything to do with deliberate falsification of reported climate records. It is frightening to realize the breadth and depth of the deception. 'Climate science' has now reached the point where supposedly trustworthy scientists and government agencies in control of temperature records have the power to make the Earth's temperature and sea level appear to be anything that they so desire. Fortunately, it is not yet too late to reverse this trend.

Chapter 10: The Reality of Global Warming

"I don't have a private jet. I live a carbon-free lifestyle."
- Al Gore[1]

Al Gore's above pronouncement is just as misleading as everything else he has said about global warming. His lifestyle can hardly be described as 'carbon-free.' While he has been careful not to actually purchase his own private jet, he borrows and charters such jets all the time to fly around the world in Gulfstream luxury.[2] The National Center for Public Policy Research reported[3] that his Tennessee home consumes 21 times the electricity than that of the average American family. The electricity used to heat his swimming pool alone would power six homes for a year.

Does Al Gore spend much time worrying about the extent to which *he* is destroying the planet? Is he concerned about how much carbon dioxide *he* spews into the upper atmosphere as he flies around the world in private jets? Does he feel guilty about the $70 million deposited into his personal bank accounts[4] from the sale of his Current TV network to Al-Jazeera (owned by the *oil-rich* nation of Qatar)? Does he lose sleep at night thinking about all of the low- and middle-income families that have been made 'energy poor' as a result of the energy policies that he espouses?

Al Gore is not a nervous wreck over climate change like many of his loyal followers because he knows that: a) the world really isn't being destroyed by global warming, and b) that he is a member of the privileged ruling elite class that will always be immune to the draconian policies that the global warming movement is inflicting on the United States and the world. His only real worry is whether he can continue to earn hundreds of millions of dollars by serving as the primary spokesman for the climate change movement.

Motivations of the Global Warming Movement

The greed exhibited by Al Gore is just one of many secondary motives driving the global warming movement. For the scientific community, another basic motivation is self-preservation. Even the most objective scientist cannot fail to recognize that: 1) most of the unelected bureaucrats in charge of federal research funding and government laboratories are now progressive socialists, 2) most of the professors at major colleges and universities in control of the peer review process are now progressive socialists, 3) most of the editors at major publishing houses and media outlets are now progressive socialists, and 4) many administrators and teachers within the public school system are now progressive socialists. The vast majority of these socialists are completely intolerant of any ideas that run counter to mainstream progressive dogma. Therefore, most scientists know that they must be careful to exhibit 'politically correct' behavior in order to survive (see Chapter 9). Anyone who dares to challenge any of the cornerstones of the global warming movement will be viciously attacked. Scientists know that if they are labeled as 'climate deniers' it will be difficult for them to obtain research funding, publish papers, or even stay employed.

The reality is that global warming is a political rather than a scientific issue and has nothing to do with 'saving the planet.'[5]

A clear demonstration of this fact is provided by the text of the 2015 Paris Agreement, which is the most far-reaching international pact on climate change ever created. This document states that climate action must include concern for "gender equality, empowerment for women, and intergenerational equity" as well as "climate justice."[6] Governments around the world are being advised that all of these steps must be implemented in order to mitigate the evils of global warming.

The primary motive behind the global warming movement is political ideology. The Cold War never ended. It is currently being fought every single minute of every single day within the boundaries of every nation on Earth. The socialist leaders of the global warming movement represent one dimension of this war. These socialists believe that the world and all people in it should be controlled by a massive all-seeing and all-knowing communistic government. To these people, capitalism and free markets are evil.

Personal freedom and democracy are evil. As the United States represents the primary embodiment of these attributes, America is an evil empire that needs to be destroyed. As America is currently too strong to be defeated militarily, its destruction must be accomplished from within.

Fossil fuels represent the primary source of affordable energy for the United States and the entire world (see Chapter 8). Fossil fuels power all democracies and free market economies, providing people with a standard of living that is unmatched in the history of the world. The industrialized world cannot function without affordable energy. Affordable energy is critical for providing heat and electricity to homes and businesses, as well as for all forms of modern transportation. The entire economic system of modern democracies would collapse without affordable energy. Unfortunately, this collapse is precisely what the worldwide socialist movement desires. This is why attacks on fossil fuels are so relentless. This is why the socialist movement is devoting so much attention to the global warming movement. They have no intention of 'saving the planet.' Their goal is to seize total control of the planet and all people in it.

It is Not Too Late to Fight Back Against 'Fake Science'

The climate change community is constantly warning people that the Earth is approaching a 'tipping point' beyond which it will be beyond saving. In reality, democratic societies may be approaching a tipping point beyond which the damage caused by the global warming movement becomes irreversible.

Fortunately, it is not yet too late to 'save the planet' from the climate change movement:

- It is fortunate that there are still scientists willing to check and challenge the many fraudulent claims made by the global warming movement. These scientists haven't yet been put in jail.
- It is fortunate that archival temperature records still exist that conclusively show that the Earth's climate has been oscillating due to natural solar cycles rather than warming due to fossil fuel

emissions. These climate records have yet to be deleted or completely replaced using massive computer programs.

- It is fortunate that we have access to satellites that can scan the entire globe to obtain accurate climate information in spite of attempts by the progressive climate change community to control, manipulate, and pervert this data (see Chapter 9).

What is unfortunate is that government agencies such as NASA and NOAA continue to block free and open access to their data and their methods. The public should demand that this information be made available to all. The public should demand that these agencies justify their handling of climate information. If these agencies refuse to supply such information even in response to either Freedom of Information Act inquiries or requests by Congressional committees (see Chapter 9), what does this say about the practices of the climate change community? Something rotten is clearly going on behind the scenes that the public and the world has a right to know.

Hopefully, the information provided in this book will provide you with a framework for evaluating climate change and its impact on democratic societies. If you wish to help preserve democracy from the further ravages of the global warming movement, the most positive things you can do are to: 1) have an open mind, 2) question what you hear, and 3) don't be afraid to challenge ideas that you think are incorrect. Look at what your children are being taught in school. Beware of media propaganda that is being passed off as factual information. Don't let emotions get in the way of logic when evaluating any information that you receive. Stop supporting or patronizing media outlets that provide propaganda rather than facts. Stop voting for politicians who hide their true agendas behind a global warming smokescreen.

The Science Really Has Been Settled

Once one moves away from political ideology and emotional appeals and gets down to cold hard logic and facts, one is forced to conclude that there is no such thing as human-induced global warming. A monumental aggregation of scientific evidence documented over hundreds of years based on disciplines including geology, chemistry, physics, biology, astronomy, meteorology, archeology, and human history fails to support the premise that fossil fuel emissions control the Earth's climate.

Much of the above evidence confirms that the temperature fluctuations experienced by the Earth in modern times are all due to natural causes such as solar cycles rather than fossil fuel emissions. Alarms about global warming are pure fiction. The falsification of climate data is absolutely essential for the survival of the global warming movement. If the climate isn't actually warming, then all of other claims of the global warming movement are exposed for the lies that they are. The Emperor has no clothes.

To summarize the data presented in this book:

- The temperature of the Earth is not spiraling out of control. Global temperatures have really only increased by 0.3°C since 1900. In fact, the current climate is cooler now than it was in 1930 prior to the era of extensive fossil fuel emissions (Chapters 1, 3, 6, and 9).
- There is no evidence that CO_2 emissions have had any impact on the Earth's temperature. Conversely, there is abundant chemical and geologic evidence pointing to the fact that existing atmospheric concentrations of CO_2 and methane should have a negligible impact on Earth's climate (Chapters 2 and 4).
- There are no hard facts to confirm that global warming is producing any dire environmental consequences including: the melting of earth's ice reserves (Chapter 5), rising sea levels (Chapter 5), extreme weather patterns (Chapter 6), or life on Earth (Chapter 7). In fact, ice levels, sea levels, weather patterns, and the diversity of life have been remarkably stable and well within levels observed throughout both human and geologic history.

From a purely factual perspective, the science of global warming is indeed settled.

Global Warming is myth.

Appendix 1: Orders of Magnitude and Scientific Notation

In dealing with climate, quantities of interest can cover many orders of magnitude. For example, events at the length scale of an atom can influence the climate on the scale of the entire planet. For this reason, scientists have developed what is called *scientific notation* to describe numbers based on factors of ten. For numbers greater than one, scientific notation involves listing factors of ten by the number of zeros that would be used in standard notation. For example, $100 = 10^2$ because there are two zeros in the number one hundred. The number 3,000 is written as 3×10^3. In these numbers, the superscript is called an order of magnitude. For numbers less than 1, the superscript refers to one less than the number of zeros to the right of the decimal point in decimal notation and is listed as a negative number. For example, 1/100 (one hundredth), which is equal to 1%, is written as 10^{-2} in scientific notation and 0.01 in standard decimal notation. Many quantities have Greek prefixes that describe their order of magnitude. For example, the prefix kilo stands for one thousand in words such as kilometer (one thousand meters). Common Greek prefixes and the other notations used to describe the same orders of magnitude are given below:

Greek Name	English Name	Order of Magnitude	Standard Notation
Tera (T)	trillion	10^{12}	1,000,000,000,000
Giga (G)	billion	10^9	1,000,000,000
Mega (M)	million	10^6	1,000,000
Kilo (k)	thousand	10^3	1,000
Centi (c)	one hundredth	10^{-2}	0.01 (1%)
Milli (m)	one thousandth	10^{-3}	0.001 (0.1%)
Micro (µ)	one millionth	10^{-6}	0.000001
Nano (n)	one billionth	10^{-9}	0.000000001

Appendix 2: Physical Quantities

Most countries and scientists use the metric system, which is based on factors of ten, to describe most physical quantities. Some countries, including America, use the English system of measure. The entries of common units listed below allow one to convert from one system of units to another. Common abbreviations are given in parentheses.

Length
1 kilometer (km) = 1,000 meters = 0.625 miles (mi)
1.6 kilometer = 1 mile
1 meter = 1.09 yards (yd) or 3.28 feet (ft)
2.54 centimeter (cm) = (1/100 m) = 1 inch (1")
1 millimeter (mm) = (1/1000 m)

Area
1 square kilometer (km^2) = 0.39 square miles (mi^2)
2.56 square kilometer (km^2) = 1 square mile (mi^2) or 640 acres
1 square meter (m^2) = 1.2 square yard (yd^2)

Volume
1 cubic kilometer (km^3) = 0.244 cubic miles (mi^3)
4.1 cubic kilometer (km^3) = 1 cubic mile (mi^3)
1 cubic meter (m^3) = (1000 liters) = 1.3 yd^3 = 35 cubic feet (ft^3) = 250 gallons (gal)
1 liter (l) = 1.06 quart (qt)

Weight (or mass)
Gigaton (Gton) = one billion metric tons
1 metric ton (1,000 kilograms) = 1.1 ton = 2,000 pounds
1 kilogram (kg) = 1,000 grams) = 2.2 pounds (lb)

Temperature

Temperatures are commonly listed in either degrees centigrade (°C) (metric) or degrees Fahrenheit (°F) (English). The centigrade scale is based on the properties of water, where 0°C is the freezing point and 100°C is the boiling point. Scientists often use temperatures in degrees Kelvin (°K), which are referenced to a temperature of *absolute zero* which is so cold that absolutely no motions can occur. The relationships between these temperature units are:

$$°K = 273 + °C$$
Room temperature = 25°C = 298°K

$$°F = 9/5 \ °C + 32$$
$$°C = 5/9 \ (°F - 32)$$
Room temperature = 25°C = 70°F

A change in temperature (ΔT) by one °C is almost twice as great (9/5) as a change of one °F.

Molecular Properties

As atoms and molecules are incredibly small, a unit called the *mole* was devised to describe the number of molecules in objects that are of human dimensions.

1 mole = 6 x 10^{23} atoms (or other objects such as molecules or photons)

For example, one liter (or quart) of water contains 55 moles of water molecules (or 3.3 x 10^{25} molecules).

All atoms and molecules have weight (or mass). This weight is typically described in terms of the weight of a mole of these tiny objects. The *molecular weight* is given in grams per mole (g/mole). Hydrogen atoms have a molecular weight of one g/mole. Carbon dioxide (CO_2) has a molecular weight of 44 g/mole, which is equal to the sum of the molecular weight of one carbon (C) atom (12 g/mole) plus two times the weight of an oxygen atom (16 g/mole).

Concentration

In liquids (such as in the ocean), concentrations are typically given in moles/liter or molar (M). In the gas phase (or the atmosphere), several different units are used. If the gas of interest has a high concentration, units used include mole percent or mole% (moles of the gas of interest divided by the moles of all gases present times 100) or weight percent (grams of the gas of interest divided by the weight of all gases present times 100).

If the gas of interest has a relatively low concentration, the concentration is described in terms of *parts per million* (ppm). The two distinct 'parts per million' units used parallel the mole percent and weight percent units described above. A part per million by volume (ppmv) corresponds to one molecule out of one million molecules present. A part per million by weight (ppmw) corresponds to one gram out of one million grams of all species present.

For example, carbon dioxide is present in the atmosphere at a concentration of 400 ppmv, which means that there are 400 molecules of CO_2 present for each million molecules in the air. The fraction of air molecules that are CO_2 is equal to $400/10^6 = 1/2,500 = 0.0004 = 0.04$ mole%.

To calculate ppmw from ppmv, one needs to know the molecular weight of the species of interest relative to the average molecular weight of the medium within which the species is dispersed. For example, air predominantly consists of nitrogen (78 mole%, with a molecular weight of 28 g/mole) and oxygen (21 mole%, with a molecular weight of 32 g/mole), giving air an average molecular weight of 28.7 g/mole. Carbon dioxide has a molecular weight of 44 g/mole. As CO_2 is 44/28.7 times heavier than air, one ppmw of CO_2 is around 1.5 times greater than one ppmv. In this specific case, 400 ppmv of CO_2 corresponds to 613 ppmw.

Appendix 3: Energy

Energy represents the most important physical quantity controlling weather and climate. Weather and climate involve the interplay and transfer of energy between light, temperature, heat, and the motion of air and water. Each of these energy components, as well as energy used by humanity, has its own unique units. Below, a brief description of the major types of energy discussed in this book is provided with examples aimed at putting each energy unit into perspective. This is followed with a table of conversion factors between the various energy units all referenced to a kilowatt-hour.

Motion

Motion requires energy. In terms of weather and climate, energy moves the air and water to create wind, waves, and oceanic currents. Energy can also be extracted from moving objects. The pistons in cars extract energy from expanding gases, while renewable energy sources such as hydroelectric power and wind energy rely on the motions of water and air to generate electricity.

The energy associated with motion is called *kinetic energy*. The energy unit used to describe kinetic energy is the *joule*, which is the standard scientific unit for energy. One joule is the energy consumed in accelerating a weight of one kilogram at a rate of 1 meter/second^2 over a distance of one meter. A joule (J) has units of kg-m^2/s^2.

Example: A ten pound (or 5 kilogram) bowling ball dropped from a height of 6 feet (or two meters) and accelerated by the force of gravity (around 10 m/s^2) gains an amount of kinetic energy of around 100 J or 0.1 kJ by the time it strikes the Earth (i.e. enough energy to smash your big toe!).

Heat

Heat describes the energy required to increase temperature. The standard unit for heat is the *calorie* (cal). A calorie is the energy needed to raise the temperature of one gram of water by one °C (around 2°F). One calorie is equal to 4.2 joules. When dieters are 'counting calories', they are actually counting kilocalories (kcal) or one thousand times the energy of a conventional calorie. Chemists often refer to the energy associated with a mole of atoms or molecules (kcal/mole). Larger quantities of heat are described in terms of a *British thermal unit* (btu), which is the energy required to raise the temperature of one pound of water by one °F.

Example: The amount of energy required to raise the temperature of a cup (around 250 ml) of water from room temperature up to near boiling (a temperature increase of around 60°C) is 15,000 calories, 15 kcal, or 63 kJ. It is perhaps surprising that this is around 630 times the energy of the dropped bowling ball in the motion example above.

Imagine how much energy it would take to raise the temperature of the entire Earth by 1°C (or 2°F) and you can begin to appreciate how much energy is required to drive the Earth's climate.

Temperature

Energy is required just to maintain the temperature of a given object. Temperature is associated with motion. At a molecular level, the kinetic energy associated with temperature is equally divided between translations (motions in three dimensions), rotations (spinning), and vibrations, where atoms connected to each other by chemical bonds oscillate back and forth. Each distinct motion involving a mole of atoms or molecules requires an amount of energy (E) that defines the effective temperature of those atoms as indicated by the equation:

$$E = \frac{1}{2} RT$$

Here, R is the ideal gas constant ($R = 2$ cal/(mole-°K) and T is the temperature in °K.

Example: Simple diatomic gases such as the nitrogen (N_2) and oxygen (O_2) in the Earth's atmosphere can move in five different ways (three translations corresponding to three dimensions, one rotation, and one vibration). This means that the energy required to maintain air at room temperature (25°C or 298°K) is (5)(1/2 RT) or 1.5 kcal/mole. As a mole of air at atmospheric pressure occupies 22.4 liters, this means that the molecules in one liter (or quart) of air are moving with a total kinetic energy of 67 calories or 280 joules (almost three times the energy of the dropped bowling ball above).

Light

The absorption of emission of visible light from matter involves the movement of electrons between what are called electronic energy levels within a given substance. If the electron moves from a high energy level to a lower energy level, light is emitted having an energy equal to the energy difference between the two levels. In light absorption, the electron absorbs the light to gain enough energy to move from a low energy to a higher energy level.

Because light absorption and emission processes involve the energies of electrons, the unit used to describe this energy is called an *electron volt* (eV). One electron volt corresponds to the energy required to accelerate an electron across an electrical potential of one volt. Because this energy is so small, the value for electron volts in the table below is for a mole (6×10^{23}, see Appendix 2) of electrons (i.e. eV/mole). One eV/mole is equivalent to 23 kcal/mole. Exact relations between the energy, frequency, and wavelength of light are provided in Appendix 4.

Example: Green light (with a wavelength of around 500 nm) is emitted with 2.5 eV of energy. A mole of green light photons contains 58 kcal (243 kJ) of energy, or enough energy to heat almost ten cups of coffee in the Heat example above.

Electricity

Electrical energy that is supplied to most homes and businesses is described in units of kilowatt-hours. A watt is a unit of power that is equal to one joule per second (J/s) or (in electrical terms) a volt-ampere. A kilowatt is one thousand watts.

Example: A kilowatt-hour is the amount of energy consumed by ten one hundred watt light bulbs over the course of one hour. Keeping a single 100-watt light bulb on for one hour consumes 360 kJ of energy, or almost 6 times the energy required to heat up a cup of coffee.

Energy Conversion Factors

1 kilowatt-hour =
3.6×10^6 joules or 3,600 kJ
8.6×10^5 calories (cal) or 860 kcal
3,413 British thermal units (btu)
1.34 horse power-hour (hp-hr)
37 electron volts per mole (eV/mole)

The above table indicates that:

A kilowatt-hour is just slightly greater than one horse power-hour and just slightly less than a thousand kilocalories (or a thousand 'food' calories, which is roughly half of the daily food intake for an average adult).

Appendix 4: Light

Fundamentals

The fundamental unit of light is a photon. Photons are considered to be particles that have no weight and no charge, but are pure energy. Light is also described as an electromagnetic wave (Fig. 2.2). Peaks in the wave correspond to maxima in electromagnetic energy. The distance between these peaks corresponds to the *wavelength* of light (given the symbol□λ). The number of peaks in the wave that move through a given point in space per unit time corresponds to the *frequency* (ν) of the light). The *amplitude* (or height) of the peaks is used to describe the intensity of the light, or the number of photons involved in the light wave. All electromagnetic waves travel at the same speed, which is the speed of light (c) (300,000 kilometers per second or 186,000 miles per second).

Relationships Between Frequency, Wavelength, and Energy

Because the speed of all frequencies of light is identical, the frequency is inversely proportional to the wavelength as given by the expression:

$\nu = c/\lambda$ (Eq. 1)

This means that an increase in frequency by a factor of ten corresponds to a decrease in wavelength by a factor of ten.

The energy of light (E) is directly proportional to its frequency and inversely proportional to its wavelength as given by the expression:

$E = h\nu$ (Eq. 2)

In this expression, h stands for Planck's constant (h = 6.6×10^{-27} erg-sec).

Units of Frequency, Wavelength, and Light Energy

Wavelengths are listed in metric units of length or distance. For example, light such as radio waves having long wavelengths are listed in meters, while wavelengths for visible light are typically given in nanometers (nm or 10^{-9} meters).

The standard unit of frequency is cycles/second (sec^{-1}) or Hertz (Hz), for the number of waves that pass a specific point per unit time. Another common frequency unit, used primarily for infrared radiation, is given by one divided by the wavelength in centimeters (1/cm or cm^{-1}). As indicated in Eq. 1, a frequency of 1 cm^{-1} is equal to 3 x 10^{10} Hz.

The standard unit used to describe the energy of a single photon of light is the electron volt (eV)(see Appendix 2). The energy of a photon having a wavelength of 1000 nm is 1.24 eV. As energy is inversely proportional to wavelength, this means that the energy of green light having a wavelength of 500 nm is around 2.5 eV. However, most energy units listed in Appendix 3 are associated with a mole of atoms (Appendix 2) rather than a single atom. For this reason, the values associated with electron volts in Appendix 3 are 6 x 10^{23} times the energy associated with a single photon or atom. As examples, a mole of 1 eV photons has the energy equivalent of 2.6 x 10^{-2} kilowatt-hours or 23 kcal/mole. A mole of green light photons contains the energy equivalent of 6.6 x 10^{-2} kW-hr (around 7% of a kilowatt-hour).

Note: Web sites are available that will calculate wavelengths, frequencies, and energies for any selected light wave. For example, see UV-Visible Spectroscopy-Michigan State University, https://www2.chemistry.msu.edu/faculty/reusch/VirtTxtJml/Spectr py/UV-Vis/spectrum.htm

Appendix 5: The Aqueous Chemistry of Carbon Dioxide

Anyone who has sampled carbonated beverages knows that carbon dioxide gas dissolves in (and can be released from) water. What many people fail to realize is that enormous amounts of carbon dioxide are dissolved in the vast oceans of Earth, and that much of the carbon dioxide released by burning fossil fuels ends up in the oceans (see Chapter 4). People may have also heard of 'carbonic acid', which is a substance that forms when carbon dioxide dissolves in water. Claims have been made that this carbonic acid is so acidic that human-derived fossil fuel emissions are starting to destroy all marine life (see Chapter 7). To evaluate this claim, one must have a basic understanding of the aqueous chemistry of carbon dioxide.

Some of the key species that control the behavior of carbon dioxide in water include:

Water = H_2O
Carbon Dioxide Gas = CO_2 (g)
Carbonic Acid = H_2CO_3
The Bicarbonate Anion = HCO_3^-
The Carbonate Anion = CO_3^{2-}
Calcium Carbonate = $CaCO_3$

People may be familiar with the bicarbonate anion in sodium bicarbonate (bicarbonate of soda) found in antacids. Carbonates are also used in antacids, and are found in the calcium carbonate that forms limestone (Chapter 4) and seashells (Chapter 7).

Equilibrium relationships between the above species allow scientists to calculate exactly how much carbon dioxide should dissolve into the oceans and how acidic the oceans should become in response to the dissolution of known quantities of CO_2. This appendix provides a brief overview of the aqueous chemistry of carbon dioxide for the interested reader. Conclusions based on this chemistry are discussed in major book chapters such as Chapters 4 and 7.

The Solubility of Carbon Dioxide Gas in Water

Henry's Law describes the relative amounts of carbon dioxide gas that are present in air and water in contact with each other at equilibrium (see Chapter 4):

CO_2 (g) + H_2O (liquid) $\longleftrightarrow$ CO_2 (aq)
k_{Henry} = 0.033 M/atm = [CO_2 (aq)]/[CO_2 (g)] (Eq. 1)

Here, CO_2 (aq) is the concentration of carbon dioxide gas dissolved in water in moles per liter (M), while the concentration of CO_2 in the air is in units of atmospheres (atm). This expression shows that the ratio of the concentrations of carbon dioxide in the air and water is always the same at a given temperature (see Chapter 4). In the context of global warming, this means that when additional carbon dioxide is introduced into the atmosphere, much of it dissolves into the oceans until the proper equilibrium ratio is reestablished. At the current atmospheric CO_2 concentration of 400 ppmv, the concentration of CO_2 (aq) in pure water is 1.3×10^{-5} M.

There are two other important factors that control just how much carbon dioxide dissolves in the oceans. First, the Henry's Law constant depends on temperature, reflecting the fact that carbon dioxide gas is less soluble in hot water than in cold water (see Chapter 4). This means that if the oceans warm up, they release more carbon dioxide into the atmosphere, while if the oceans cool, they absorb more CO_2 from the air.

The second important factor controlling carbon dioxide dissolution is its conversion into other chemical species. These conversions consume CO_2 (aq), which allows more carbon dioxide gas to dissolve in order to replace the CO_2 (aq) that is lost. Equilibrium constants between all products formed from carbon dioxide must be taken into account in order to actually calculate how much carbon dioxide is removed from the air by the oceans. Equilibrium constant expressions involving the major species derived from carbon dioxide are described below following a brief discussion of acid-base chemistry.

A Primer on Acids and Bases

In the context of this book, an acid (called a Bronsted acid) is any substance that can dissociate (or fall apart) to donate protons (H^+ or more accurately hydronium ions (H_3O^+)) to an aqueous solution. A base is any substance that donates hydroxide ions (OH^-) to solution. A water molecule can donate both species to solution as indicated by the reaction:

$$H_2O \leftrightarrow H^+ + OH^- \quad k_w = [H^+][OH^-] = 10^{-14} \, M^2 \quad (Eq. 2)$$

This expression shows that in pure water, $[H^+] = [OH^-] = 10^{-7}$ M. As the concentration of water molecules in liquid water is 55 M, only one out of every 550 million water molecules dissociates, making the concentration of protons in pure water very low. If protons are added to solution, hydroxide concentrations must drop to compensate. For example, if acid is added to make $[H^+] = 10^{-4}$ M, the hydroxide concentration must drop to $[OH^-] = 10^{-10}$ M to satisfy Eq. 2.

The proton concentration in any solution is normally described in terms of the solution pH as defined by:

$$pH = - \log [H^+] \quad (Eq. 3)$$

Here, the log of the concentration is expressed as a power of ten. For example, the pH of pure water is 7, corresponding to a concentration of 10^{-7} M. Solutions having a pH that is lower than 7 (e.g. pH 2) are considered to be acidic, while those having a pH greater than 7 (such as the pH of 8.2 found in seawater) are basic.

The strength of an acid is defined by the extent to which a given substance dissociates to release protons as defined by its *acid-base equilibrium constant*. When a strong acid such as hydrochloric acid is introduced into water, it dissociates completely via reactions such as:

$$HCl \leftrightarrow H^+ + Cl^- \quad (Eq. 4)$$

This means that the concentration of protons is identical to the concentration of HCl that was added to the solution. However, a weak acid does not dissociate completely. For example, the acetic acid (CH_3COOH) found in vinegar dissociates according to the equilibrium expression:

$$CH_3COOH \longleftrightarrow H^+ + CH_3COO^-$$
$$K_a = [H^+][CH_3COO^-]/[CH_3COOH] = 1.8 \times 10^{-5} \text{ M} \quad \text{(Eq. 5)}$$

The proton concentration in a 1 M solution of acetic acid is only 4×10^{-3} M (250 times lower than that in 1 M HCl) resulting in a pH of 2.4. While vinegar is considerably more acidic than carbonic acid (see below), it is still safe to eat on salads. Given this context, one can now evaluate the dangers associated with the acidity of solutions containing dissolved carbon dioxide.

The Formation and Acid-Base Chemistry of Carbonic Acid

Roughly 0.1% (1/1000) of the carbon dioxide molecules entering an aqueous solution can react with water (H_2O) molecules to form molecules of *carbonic acid* (H_2CO_3) by the reaction:

$$CO_2 \text{ (aq)} + H_2O \text{ (l)} \longleftrightarrow H_2CO_3$$
$$K_{hyd} = [H_2CO_3]/[CO_2 \text{ (aq)}] = 1.6 \times 10^{-3} \quad \text{(Eq. 6)}$$

Carbonic acid is a diprotic acid, which means that it has the potential to release two protons (H^+) into solution. Loss of the first proton is described by the following acid-base reaction:

$$H_2CO_3 \longleftrightarrow H^+ + HCO_3^-$$
$$K_{a1} = 3.2 \times 10^{-4} \text{ M} = [H^+][HCO_3^-]/[H_2CO_3] \quad \text{(Eq. 7)}$$

Although is it not technically correct to do so, CO_2 (aq) is often described as being equivalent to carbonic acid by combining equations 6 and 7 to get:

$$CO_2 \text{ (aq)} + H_2O \longleftrightarrow H^+ + HCO_3^-$$
$$K_{a1*} = [H^+][HCO_3^-]/[CO_2 \text{ (aq)}] = 4.2 \times 10^{-7} \quad \text{(Eq. 8)}$$

212

Using the above expression in combination with Henry's Law, one can calculate the pH of pure water in contact with any atmospheric concentration of carbon dioxide. These calculations clearly show that any claims regarding environmental catastrophes due to carbon dioxide-induced acid rain are completely unfounded (see Chapter 7). At existing carbon dioxide levels, rainwater has a pH of 5.63 ($[H^+]$ = 2.3 x 10^{-6} M). *Even if* all fossil fuels on Earth were incinerated and *even if* all of the resulting CO_2 remained in the atmosphere (which it doesn't), the resulting pH of rainwater would only be lowered to 5.41.

The Acid-Base Chemistry of the Bicarbonate Ion

While rainwater is relatively pure, the water in the oceans is not. Seawater contains a wide range of salts and other dissolved inorganic substances in addition to carbon dioxide. As a result, the ocean is basic rather than acidic, with a pH of 8.2. In seawater, essentially all carbonic acid has been converted into the bicarbonate ion. In fact, the ocean is sufficiently basic that bicarbonate ions can lose protons to form carbonate ions (CO_3^{2-}) via the reaction:

$HCO_3^- \longleftrightarrow H^+ + CO_3^{2-}$
$K_{a2} = [H^+][CO_3^{2-}]/[HCO_3^-] = 4.8 \times 10^{-11}$ (Eq. 9)

The formation of the carbonate ion consumes some of the bicarbonate, which in turn consumes some of the carbonic acid in solution, which in turn allows further atmospheric carbon dioxide to dissolve into the ocean.

The amounts of carbonic acid, the bicarbonate ion, and the carbonate ion in aqueous solutions in contact with the Earth's current atmosphere (400 ppmv of CO_2 (g)) are shown in the Appendix 5 figure below versus solution pH. The dark curve in the figure depicts the total amount of dissolved carbon-containing species that are derived from carbon dioxide, while the lighter straight lines represent the concentrations of carbonic acid, the bicarbonate ion, and the carbonate ion on a logarithmic (log) scale (i.e. each number represents ten times the concentration of the number beneath it). The dark curve clearly shows that above a pH of 6, the total concentration starts to show a dramatic increase as carbonic acid starts to be converted into the bicarbonate ion. *Due to bicarbonate formation, seawater (at pH 8.2) dissolves almost one hundred times more CO_2 (a total of 2.4×10^{-3} M) than pure rainwater does.* Above pH 10, there is a further increase as the bicarbonate ion starts to be converted into the carbonate ion.

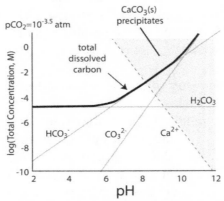

Appendix 5 Fig. Concentrations of carbonic acid, the bicarbonate ion, the carbonate ion, and the total concentration of all CO_2-based species as a function of solution pH. (Reprinted from Bunker, B. C. and Casey, W. H., The Aqueous Chemistry of Oxides, Oxford University Press, New York, 2016.

Carbonate Precipitation

The ultimate fate of dissolved carbon dioxide involves the formation of insoluble carbonate precipitates. Solid carbonates such as the calcium carbonate that comprises limestone form via reactions such as:

$$Ca^{2+} + CO_3^{2-} \longleftrightarrow CaCO_3$$
$$K_{sp} = 1/([Ca^{2+}][CO_3^{2-}]) = 5 \times 10^7 \text{ M}^{-2} \qquad \text{(Eq. 10)}$$

Precipitation removes the carbonate ion from solution, allowing more carbon dioxide to dissolve into the ocean. Eq. 10 shows that calcium carbonate can be made to either dissolve or precipitate in response to changes in solution chemistry. A lowering of either calcium or carbonate concentrations can stimulate dissolution, while increases in the concentration of either species can promote precipitation. Because protons react with the carbonate ion to form the bicarbonate ion, one factor that can lower carbonate concentrations is pH. For this reason, at a fixed atmospheric CO_2 concentration, calcium carbonate is more soluble in acidic solutions than in basic solutions, as indicated by the gray region in Fig. 5A.

The pH of Seawater: Dissolved Carbonates as pH Buffers

Anyone who has suffered from acid indigestion or heartburn knows that common ingredients in antacid tablets are bicarbonate of soda or either calcium or magnesium carbonate. This is because both substances are capable of neutralizing rather than creating acid. In fact, solutions containing mixtures of carbonic acid, sodium bicarbonate, and sodium carbonate are *pH buffer solutions*. A buffer solution is one that is resistant to changes in pH. Some of these buffers are used in aquariums to stabilize the pH in the tank to stay fixed near desired values. For example, when acid is added to a solution containing carbonate (CO_3^{2-}) ions, the protons in the acid are consumed to form bicarbonate ions (HCO_3^-), resulting in little change in pH. Instead, what changes are the relative amounts of carbonate and bicarbonate ions. The ability of dissolved bicarbonate and carbonate ions to buffer pH is one reason why seawater is less susceptible to pH changes than fresh water.

The Net Impact of Dissolved Carbon Dioxide on Oceanic pH

As the above discussion indicates, the chemistry of carbon dioxide in water is complex. In order to determine the impact of dissolved carbon dioxide on pH, all of the chemical equilibria listed in Eq. 6-10 (and a wide range of equilibria involving other dissolved species) must be taken into account. Geochemists have developed computer codes that can simultaneously solve all of these interrelated equilibrium expressions to provide concentrations of all species including protons and hydroxide ions. A few of these programs are available through U.S. government agencies and universities (Chapter 7, Ref. 39). These programs have been used to calculate that:

The minimum pH that would be produced in the oceans if all fossil fuels on Earth were to be burned is pH 8.0 (see Chapter 7).

Appendix 6: The Impact of Melting Icebergs, Pack Ice, and Shelf Ice on Sea Levels

An iceberg forms when a massive chunk of ice calves off a coastal glacier and falls into the ocean. If ice were heavier than water, the iceberg would sink to the bottom and add its entire volume to that of the ocean. However, ice (with a density of 0.92 gm/cm^3) is 8% lighter than water (whose density is 1.0 gm/cm^3). This means that ice is a *buoyant* object that floats on water. The fraction of the ice volume that is under water is equal to the ratio of the density of ice to that of water. This means that 92% of the iceberg is under water, while 8% protrudes above the surface. This also means that only 92% of the iceberg's total volume is added to that of the ocean.

What happens when this iceberg melts? In terms of a further sea level change, the answer is nothing. The part of the iceberg that is underwater melts to form water. However, as water is 8% denser than ice, this means that the melting process actually reduces the net volume that was occupied by the ice. The melting of the ice that was above the ocean surface exactly compensates for this volume reduction. This means that the melting of the iceberg results in no further change in sea level.

If you do not believe that this is true, conduct your own 'sea level' experiment. Take a clear glass and fill it half full with water. Add all of the ice that you wish so long as no ice touches the bottom of the glass. Make a mark on the glass corresponding to the level of the water in the glass. Now let the ice melt. You will find that there is no change in the 'sea level' in your glass even when all of the ice is gone.

The above experiment is critical to understanding what happens when pack ice and shelf ice melt. Both ice forms start out as objects that are floating in the ocean. Therefore, when either object melts, it contributes absolutely nothing to the sea level.

Some environmentalists try to argue that this is an oversimplification because salty seawater is 2-3% denser than the freshwater released by a melting iceberg. Therefore, they claim that 2-3% of the volume of either pack or shelf ice that melts contributes to rising sea levels. However, this ignores the fact that the fresh water released by any ice in the ocean almost always immediately mixes with ocean water, is diluted with salt water, and actually exhibits no density change whatsoever. Even if one assumes that there is absolutely no mixing between iceberg melt-water and the ocean, the maximum possible contributions that the melting of all pack and sea ice on Earth could make to sea levels are 3% of the values discussed in Chapter 5 and Table 5.1:

Total Melting of All Pack Ice = 3% of 10 cm = 3 millimeters
Total Melting of All Shelf Ice = 3% of 35 cm = 1 centimeter

References

Chapter 1

1. Lenin on lying, www.quotationspage.com/quote/838.html

2. Watts, A., Global Cooling Compilation, March 1, 2013, htpps://wattsupwiththat.com/2013/01/01/global-cooling-compilation/

3. Gwnne, P., Newsweek, Apr. 28, 1975, p. 64

4. The Big Freeze (cover story), Time Magazine, Jan. 31, 1977, p. 86

5. Chapman, M. W., "Transcript: Al Gore Got D in Natural Sciences at Harvard," May 24, 2011, www.cnsnews.com/news/article/transcript-al-gore-got-d-natural-sciences-harvard

6. Revelle, R., and Suess, H., "Carbon Dioxide Exchange Between Atmosphere and Ocean and the Question of an Increase in Atmospheric CO_2 During the Past Decades," Tellus, 9, 18-27 (1957)

7. Sussman, B., Eco-Tyranny, WND Books, Washington, D.C., 2012, p. 59

8. Gore, A., "To Skeptics on Global Warming," New York Times, April 22, 1990

9. Gore, A., Earth in the Balance: Ecology and the Human Spirit, Houghton Mifflin, 1992

10. Al Gore, blog entry, "Fox News Manipulates Climate Coverage," Dec. 15, 2010

11. KUSI News Report (San Diego), Oct. 2, 2010 at www.kusi.com/story/13257617/roger-revelle-al-gore-colemans-video-report-3609

12. President's Council on Sustainable Development, Revised Charter, Scope of Activities, April 25, 1997

13. Savage, M., Government Zero, Utopia Publications, Inc., New York, 2015, p. 232

14. Gore, A., An Inconvenient Truth: The Planetary Emergency of Global Warming and What We Can Do About It, Rodale Press, Inc. 2006

15. Savage, M., Government Zero, p. 216, Utopia Publications, Inc., New York, 2015

16. Bowden, T. A., Marland, G., and Andres, R. J., Global, Regional, and National Fossil Fuel CO_2 Emissions, Carbon Dioxide Information Analysis Center, Oak Ridge National Laboratory, U. S. Department of Energy, Oak Ridge, Tenn., U.S.A. doi 10.3334/CDIAC/00001 (2017)

17. Burton, Dave, http://www.sealevel.info/CO2_and_CH4.html

18. Mann, M. E., Bradley, R. S., Hughes, M. K., "Global-Scale Temperature Patterns and Climate Forcing Over the Past Six Centuries," Nature, 392, 779-787 (1998)

19. International Panel on Climate Change Third Assessment Report, 2001, p. 29, Fig. 2.3

20. Arter, M., "Obama Links Islamic Terrorism to Climate Change," cnsnews.com, May 20, 2015, http://www.cnsnews.com/news/article/melanie-hunter/obama-links-islamic-terrorism-climate-change.

21. Houghton, J. T., Jenkins, G. J., and Ephraums, J. J., Eds., Climate Change: The IPCC Scientific Assessment, Cambridge University Press, Cambridge, Great Britain, 1990, p. 202, Fig. 7.1c.

22. Monckton, C., Climategate: Caught Green-Handed, Science and Public Policy Institute Original Paper, scienceandpublicpolicy.org, Dec. 7, 2009

23. Meyer, A., "New EPA Regs Issued Under Obama are 38 Times as Long as the Bible," cnsnews.com, June 23, 2014, http://www.cnsnews.com/news/article/ali-meyer/new-epa-regs-issued-under-obama-38-times-as-long-as-bible.

Chapter 2

1. Burt, C. C., Extreme Weather, Chapters 1 and 2, p. 211, W. W. Norton & Company, Inc., New York, 2004; Photographs taken by the National Oceanic and Atmospheric Administration.

2. Bunker, B. C., and Casey, W. H., The Aqueous Chemistry of Oxides, Oxford University Press, 2016, p. 373

3. Rohde, R. A., of Global Warming Art, and Wikipedia, htt://www.globalwarmingart.com/wiki/File:Solar-Spectrum-png.

4. Primer on Solar Energy, Solar Energy Systems, Argonne National Laboratory, web.anl.gov.

5. Earth's Energy Budget, https://en.wikipedia.org/wiki/Earth's_energy_budget

6. Drago, R. S., Physical Methods in Inorganic Chemistry, Reinhold Publishing Corp., New York, 1965, p. 127.

7. a) Venyaminov, S. Y., Prendergast, F. G., "Water (H_2O and D_2O) Molar Absorptivity in the 1000-4000 cm-1 Range and Quantitative Infrared Spectroscopy of Aqueous Solutions," Anal. Biochem., 248, pp. 234-245 (1997), b)

http://www.pnl.gov/science/images/highlights/cmsd/watermodellg.j
pg.

8. Falk, M., and Miller, A. G., "Infrared Spectrum of Carbon
Dioxide in Aqueous Solution," Vibrational Spectroscopy, 4, pp.
105-108 (1992),

9. Infrared Spectra of Methane and Carbon Dioxide,
http://bitm.org/utils/common/img/chem/CO2+CH4.gif.

10. Graph of the History of Earth's Humidity (data from NOAA
Earth System Research Laboratory),
https://www.friendsofscience.org/index.php?id=710.

11. Thompson, G., The AGW Smoking Gun, American Thinker,
Feb. 17, 2010,
http://www.americanthinker.com/printpage/?url=http//www.americ
anthinker.com.

12. Beiser, A., The Earth, Chapter 3 – Anatomy of the Skies, Life
Nature Library, Time, Inc., New York, 1963, pp. 57-80.

13. Weast, R. C., Ed., CRC Handbook of Chemistry and Physics,
58th Ed., CRC Press Inc., Cleveland, Ohio, 1977, p. F208

Chapter 3

1. Adapted from data in: a) Scotese, C. R., Analysis of the
Temperature Oscillations in Geologic Eras (2002), Scotese, C. R.,
2002, http://www.scortese.com (PALEOMAP website). b)
Ruddiman, W. F., Earth's Climate: Past and Future, W. H.
Freeman & Sons, New York, 2001

2. Gould, S. J., Ed., The Book of Life, W. W. Norton & Company,
Inc., New York, 1993, pp. 25-28

3. Shields-Zhou, G. A., Hill, A. C., Macgabhann, B. A., "The
Cryogenian Period," Chapter 17 in The Geologic Time Scale, V1,

Gradstein, F. M., Ogg, J. G., Schmitz, M. D., and Ogg, G. M., Eds., Elsevier, Amsterdam, 2012

4. Gould, S. J., Ed., The Book of Life, W. W. Norton & Company, Inc., New York, 1993, pp. 46-53

5. Gould, S. J., Ed., The Book of Life, W. W. Norton & Company, Inc., New York, 1993, pp. 185-192

6. Gould, S. J., Ed., The Book of Life, W. W. Norton & Company, Inc., New York, 1993, pp. 208-210

7. Petit, J. R., et al., "Climate and Atmospheric History of the Past 420,000 Years from the Vostok Ice Core, Antarctica, Nature, 399, 429-436 (1999)

8. Mudelsee, M., The Phase Relations Among Atmospheric CO_2 Content, Temperature, and Global Ice Volume Over the Past 420 ka. Quaternary Science Reviews, 20, 583-589 (2001)

9. Frazier, K., Our Turbulent Sun, Chapter 9: Dance of the Orbits, Prentice-Hall, Inc., Englewood Cliffs, N. J., 1982, pp. 119-137

10. Dansgaard, W., et al., "Evidence for General Instability of Past Climate from a 250-kyr Ice-Core Record," Nature, 364, 218-220 (1993)

11. Bjornsson, H., The Glaciers of Iceland: A Historical, Cultural, and Scientific Overview, Atlantis Press, 2016, p. 561

12. Jones, G., A History of the Vikings, Oxford University Press, New York, 1984, p. 277

13. Jones, G., A History of the Vikings, Oxford University Press, New York, 1984, pp. 307-308

14. Frazier, K., Our Turbulent Sun, Prentice-Hall, Inc., Englewood Cliffs, N. J., 1982, p. 6

15. Frazier, K., Our Turbulent Sun, Chapter 3: The Inconsistent Sun, Prentice-Hall, Inc., Englewood Cliffs, N. J., 1982, pp. 31-48

16. Frazier, K., Our Turbulent Sun, Chapter 2: The Sun's Impact on Earth, Prentice-Hall, Inc., Englewood Cliffs, N. J., 1982, pp. 12-30

17. Frazier, K., Our Turbulent Sun, Prentice-Hall, Inc., Englewood Cliffs, N. J., 1982, pp. 36-39

18. McInnes, L. for Wikimedia Commons, Feb. 28, 2007, https://commons.wikimedia.org/wiki/File:Carbon14_with_activity_labels.svg

19. Average Climate over North America since 1880, pubs.giss.nasa.gov/docs/1999/1999 Hansen etal 1.pdf

20. Total Solar Irradiance Data compiled by the National Oceanic and Atmospheric Administration (NOAA), https://www.ngdc.noaa.gov/stp/solar/solarirrad.html.

21. Williams, M., "What Is the Life Cycle of the Sun?", universetoday.com, Sept. 24, 2016

22. Savino, J., and Jones, M. D., Supervolcano, The Career Press, Inc., Franklin Lakes, N. J. (2007), pp. 68, 70-71.

23. Renne, P. R., Zhang, Z., Richards, M. A., Black, M. T., and Basu, A. R., "Synchrony and Causal Relations Between Permian-Triassic Boundary Crises and Siberian Flood Volcanism," Science, 269,1413-1416 (1995)

24. Courtillot, V., Besse, J., Vandamme, D., Mintigny, R., Jaeger, J.-J., and Capetta, H., "Deccan Flood Basalts at the Cretaceous-Tertiary Boundary?", Earth and Planetary Science Letters, 80, 361-374 (1986)

25. Alvarez, W., T. Rex and the Crater of Doom, Princeton University Press, New York, 1997

Chapter 4

1. Bowden, T. A., Marland, G., and Andres, R. J., Global, Regional, and National Fossil Fuel CO_2 Emissions, Carbon Dioxide Information Analysis Center, Oak Ridge National Laboratory, U. S. Department of Energy, Oak Ridge, Tenn., U.S.A. doi 10.3334/CDIAC/00001 (2017)

2. Sussman, B., Climategate, WND Books, Washington, D. C., 2010, p. 78

3. Burton, Dave, http://www.sealevel.info/CO2_and_CH4.html

4. Dismukes, G. C., Klimov, V. V., Kozlov, Y. N., DasGupta, J., and Tyryshkin, A., "The Origin of Atmospheric Oxygen on Earth: The Innovation of Oxygenic Photosynthesis," Proc. Natl. Acad., Sci., 95(5), 2170-2175 (2001)

5. Gould, S. J., Ed., The Book of Life, W. W. Norton & Company, Inc., New York, 1993, p. 44

6. Adapted from data in: a) Scotese, C. R., Analysis of the Temperature Oscillations in Geologic Eras (2002), Scotese, C. R., 2002, http://www.scortese.com (PALEOMAP website). b) Ruddiman, W. F., Earth's Climate: Past and Future, W. H. Freeman & Sons, New York, 2001, c) Pagani, M., et al., "Marked Decline in Atmospheric Carbon Dioxide Concentrations During the Paleocene, Science, 309 (5734), 600-603, July 22, 2005.

7. Mudelsee, M., The Phase Relations Among Atmospheric CO_2 Content, Temperature, and Global Ice Volume Over the Past 420 ka. Quaternary Science Reviews, 20, 583-589 (2001)

8. Redd, N. T., "Venus' Atmosphere: Composition, Climate, and Weather," Space.com, Nov. 16, 2012, https://www.space.com/18527-venus-atmosphere.html.

9. "Global Oil Reserves and Fossil Fuel Consumption," The Guardian Datablog, Sept. 2, 2009, https://www.the guardian.com/environment/datablog/2009/sep/02/oil-reserves.

10. Weast, R. C., Ed., CRC Handbook of Chemistry and Physics, 58th Edition, CRC Press, Inc., Cleveland, Ohio, 1974

11. www.marinebio.net/marinescience/02ocean/swcomposition.htm.

12. Palmer, D. A., Van Eldick, R., "The Chemistry of Metal Carbonato and Carbon Dioxide Complexes," Chem. Rev., 83, 651-731, 1983

13. "What is the Carbon Dioxide Content of a Soda Can or Bottle?", https://chemistry.stackexchange.com/questions/9067/what-is-the-carbon-dioxide-content-of-a-soda-can-or-bottle.

14. For daily records, see: World Sea Temperatures, https://www.seatemperatures.org.

15. Thermocline, Oceanography, Encyclopedia Britannica, https://www.britannica.com/science/thermocline.

16. Mudelsee, M., The Phase Relations Among Atmospheric CO_2 Content, Temperature, and Global Ice Volume Over the Past 420 ka. Quaternary Science Reviews, 20, 583-589 (2001)

17. Sussman, B., Climategate, WND Books, Washington, D. C., 2010, pp. 74-75.

18. Frazier, K., Our Turbulent Sun, Chapter 9: Dance of the Orbits, pp. 119-137, Prentice-Hall, Inc., Englewood Cliffs, N. J., 1982

19. "How Much CO_2 Does a Human Exhale?", https://www.reference.com/science/much-co2-human-exhale-3f8cfdd9076c129.

20. Current World Population, worldometers.info/world-population/

21. Cattle Carbon Footprint,
http://www.americancatttlemen.com/articles/catt-carbon-footprint.

22. Raich, J., Potter, C., "Global Patterns of Carbon Dioxide Emissions from Soils," Global Biogeochemical Cycles, 9, pp. 23-36 (1995).

23. Soil Respiration, Wikipedia,
https://en.wikipedia.org/wiki/Soil_respiration.

24. Geider, R. J., et al., "Primary Productivity of Planet Earth: Biological Determinants and Physical Constraints in Terrestrial and Aquatic Habitats, Global Change Biol., 7(8), 849-882 (2001).

25. Butler, R., Global Deforestation Rates, Tropical Rainforests, Mongabay, http://rainforests.mongabay.com/deforestation.html.

26. Berner, E. K., and Berner, R. A., Global Environment: Water, Air, and Geochemical Cycles, Prentice Hall, New York, 1996.

27. Gould, S. J., Ed., The Book of Life, W. W. Norton & Company, Inc., New York, 1993, p. 43

28. Savage, S., "Volcano Output Could Be 150-300,000 Tons Daily," Red Orbit, Apr. 20, 2010,
www.redorbit.com/news/science/1852153/

29. Volcanic Gases, Oregon State University,
volcano.oregonstate.edu/book/export/html/151.

30. Savino, J., and Jones, M. D., Supervolcano, The Career Press, Inc., Franklin Lakes, N. J. (2007)

31. Understanding Plate Motions, U. S. Geological Survey,
https://pubs.usgs.gov/dynamic/understanding/html.

32. Volcanoes Can Affect the Earth's Climate, U.S. Geological Survey, https://volcanos.usgs.gov/vhp/gas_climate.html.

33. Renne, P. R., Zhang, Z., Richards, M. A., Black, M. T., and Basu, A. R., "Synchrony and Causal Relations Between Permian-Triassic Boundary Crises and Siberian Flood Volcanism," Science, <u>269</u>, 1413-1416 (1995)

Chapter 5

1. Bastasch, M., "The Arctic Still Isn't Ice Free Despite Alarmism," The Daily Caller, July 8, 2015, www.dailycaller.com/2015/07/08/the-arctic-still-isnt-ice-free-despite-alarmism/

2. Unless otherwise noted, data on ice quantities and distributions in this chapter are compiled from the National Snow and Ice Data Center, http://nsidc.org.

3. Arctic Sea Ice Extent and Antarctic Sea Ice Extent, in Chartic Interactive Sea Ice Graph, Arctic Sea Ice News and Analysis, National Snow and Ice Data Center, http://nsidc.org/arcticseaicenews/chartic-interactive-sea-ice-graph/

4. Poore, R. Z., Williams, Jr., R. S., and Tracy, C., "Sea Level and Climate," USGS Fact Sheet 0002-00, pubs.usgs.gov/fs/fs2-00/

5. Rafferty, J. P., "Cryogenian Period," Encyclopedia Britannica, https://www.britannica.com/science/Cryogenian-Period.

6. Gornitz, V., "Sea Level Rise, After the Ice Melted and Today," NASA Science Briefs, Jan. 2007, https://www.glss.nana.gov/research/briefs/gornitz_09/

7. Rohde, R. A., Post-Glacial Sea Level Rise, Global Warming Art project, https://en.wikipedia.org/wiki/Sea_level_rise#/media/File:Post-Glacial-Sea-Level.png.

8. Tilsey, P., Fox News Science, Dec. 30, 2013, "'Stuck in Our Own Experiment': Leader of Trapped Team Insists Polar Ice is Melting," www.foxnews.com/science/2013/12/30.

9. Chapman, W., "Why the Arctic is Climate Change's Canary in the Coal Mine," https://ed.ted.com/essons/

10. Rose, D., "Myth of Arctic Meltdown: Stunning Satellite Images Show Summer Ice Cap is Thicker and Covers 1.7 Million Square Kilometers More than Two Years Ago . . . Despite Al Gore's Prediction That It Would Be Ice Free by Now," Daily Mail.com, Aug. 30, 2014, www.dailymail.co.uk/sciencetech/article-2738653/

11. Ross Ice Shelf, Encyclopedia Britannica, https://britannica.com/place/Ross-Ice-Shelf, Ronne Ice Shelf, Encyclopedia Britannica, https://britannica.com/place/Ronne-Ice-Shelf

12. Pappas, S., "New Weakness in Antarctic Ice Sheet Discovered," Live Science, May 9, 2012, https://www.livescience.com/20190-weakness-antarctic-ice-sheet-melting.html

13. Sheppard, M., "Antarctica and the Myth of Deadly Rising Seas," American Thinker, Jan. 14, 2010, http://www.americanthinker.com/2010/01/antarctica_and_the_myth_of_dea_1.html.

14. Iceberg B-15, Ross Ice Shelf, Antarctica, NASA Earth Observatory, Apr. 14, 2000, https://earthobservatory.nasa.gov/OTD/view.php?id=552/

15. Carrington, D., "IPCC Officials Admit Mistake Over Melting Himilayan Glaciers," The Guardian, Jan. 20, 2010, https://www.theguardian.com/environment/2010/jan/20/

16. Bastache, M., "An Inconvenient Review: After 10 Years Al Gore's Film is Still Alarmingly Inaccurate," The Daily Caller, May 3, 2016, dailycaller.com/2016/05/03/

17. Moran, R., "Ooops! Scientists Goof on Himalayan Glacier Retreat," American Thinker, Jan. 17, 2010, http://www.americanthinker.com/blob/2010/01/ooops_scientists_goof_on_himil.html

18. Ihucha, A., "Mount Kilimanjaro Glaciers Nowhere Near Extinction," quoting Mount Kilimanjaro National Park ecologist Imani Kikoti, eTurbo News, April 6, 2014, https://eturbonews.com/44420/

19. Williams, Jr., R. S., Ferrigno, J. G., Eds., "Satellite Image Atlas of Glaciers of the World," U. S. Geological Survey Professional Paper 1386, 2010, https://pubs.usgs.gov/pp/p1386

20. World's 7 Largest Glaciers by Continent, Green Packs, Apr. 17, 2009, www.greenpacks.org/2009/04/17/worlds-7-largest-glaciers-by-continent

21. Scherler, D., Bookhagen, B., and Strecker, M., "Spatially Variable Response of Himilayan Glaciers to Climate Change Affected by Debris Cover," Nature Geoscience, 4, pp. 156-159 (2011)

22. Bjornsson, H., The Glaciers of Iceland: A Historical, Cultural, and Scientific Overview, Atlantis Press, 2016

23. Jones, G., A History of the Vikings, Oxford University Press, New York, 1984, pp. 290-293.

24. NASA – Satellites See Unprecedented Greenland Ice Sheet Melting," July 24, 2012, https://www.NASA.gov/topics/earth/features/greenland-melt.html

25. Langen, P. L., "Current Surface Mass Budget of the Greenland Ice Sheet," Danish Meterological Institute, April 25, 2017, http://www.dml.dk/en/greenland/maalinger/greenland-ice-sheet-surface-mass-budget/

26. Antarctica Climate and Weather, Cool Antarctica, http://www.coolantarctica.com/Antarctica fact file/antarctica environment/climate_weather.php.

27. Original data is from eqrthobservatory.NASA.gov/Antarctic_temps_AVH1982-2004jpg. Discussion is provided at icecap.us/images/uploads/misleading_reports_about_antarctica.pdf.

28. Zwally, H. J., Li, J., Robbins, J. W., Saba, J. L., Yi, D., and Brenner, A. C., "Mass Gains of the Antarctic Ice Sheet Exceed Losses," J. Glaciology, 61 (230), 1019-1036 (2015)

29. Gore, A., Inconvenient Truth: The Planetary Emergency of Global Warming and What We Can Do About It, Rodale Press, Inc. 2006

30. Gregory, J., Projections of Sea Level Rise, Chapter 13: Sea Level Change in: Climate Change 2013: The Physical Science Basis, https://www.ipcc.ch/pdf/unfccc/cop19/3_gregory13sbsta.pdf.

31. Poore, R. Z., Williams, R. S., and Tracey, C., Sea Level and Climate, USGS Fact Sheet 002-00, https://pubs.usgs.gov/fs/fs2-00/.

32. Rohde, R. A., Recent Sea Level Rise, Global Warming Art Project, https://commons.wikimedia.org/File:Recent_Sea_Level_Rise.png.

33. Weast, R. C., Ed., CRC Handbook of Chemistry and Physics, 58th Edition, CRC Press, Inc., Cleveland, Ohio, 1974, p. F6

34. Heller, Tony, "Falling Sea Level," March 23, 2016, https://realclimatescience.com/2016/03/falling_sea_level/

35. Mitchell, A., "Climate Change Makes Sea Levels Fall, not Rise, New NASA Study Shows," Feb. 16, 2016, https://www.christiantoday.com/article/climate.change.makes.sea.levels.fall.not.rise/79711.htm.

Chapter 6

1. Gelbspan, R., "Katrina's Real Name," Op-Ed posted Aug. 30, 2005 on
www.heatisonline.org/contentserver/objecthandlers/index.cfm?id=method=full

2. Burt, C. C., Extreme Weather, Chapters 1 and 2, pp. 15-69, W. W. Norton & Company, Inc., New York, 2004

3. htpp://www.globalwarming.org/wp-content/uploads/2012/08/Christie-Number-State-High-Low-Temperatures-Aug-2012

4. Heller, T., "The History of NASA/NOAA Temperature Corruption," p. 16, posted at
https://realclimatescience.com/2016/01/the-history-of-nasanoaa-temperature corruption/

5. Hansen, J., Reedy, R., Sato, M., and Lo, K., "Global Temperature Change", Reviews of Geophysics, 48, RG4004 (2010), Fig. 19b, p. 21 or
https://pubs.giss.nana.gov/docs/2010/2010_Hansen_ha00510u.pdf

6.
www.remss.com/data/msu/monthly_time_series/RSS_Monthly_MSU_AMSU_Channel_TLT_Anomalies_Land_and_Ocean_v03_3.txt

7. "Argo: How Argo Floats Work,",
http://www.argo.ucsd.edu/How_Argo_floats.html.

8. "Ocean Temperature," Jonova,
http://jonova.s3.amazonaws.com/graphs/ocean/global-ocean-temperature-700m-models-argo.gif.

9. Hurricane Archive,
https://wunderground.com/hurricane/hurrarchive.asp

10. Gelbspan, R., "Katrina's Real Name," Op-Ed posted Aug. 30, 2005 on
www.heatisonline.org/contentserver/objecthandlers/index.cfm?id=method=full

11. Richardson, V., "Climate Scientist Rebuts Hollywood Hurricane Hype: 'This is What Weather Looks Like', Washington Times, Sept. 21, 2017,
http://amp.washingtontimes.com/news/2017/climate-scientist-rebuts-hollywood-hurricane-hype.

12. Hurricane Archive,
https://wunderground.com/hurricane/hurrarchive.asp

13. Burt, C. C., Extreme Weather, Chapter 6, pp. 164-199, W. W. Norton & Company, Inc., New York, 2004

14. Burt, C. C., Extreme Weather, Chapter 6, pp. 164-199, W. W. Norton & Company, Inc., New York, 2004

15. U. S. Annual Count of Strong to Violent Tornados, NOAA/NWS Storm Prediction Center,
www1.ncdc.noaa.gov/pub/data/cmb/images/tornado/clim/EF3-EF5.png

16. Burt, C. C., Extreme Weather, Chapter 4, pp. 103-133, W. W. Norton & Company, Inc., New York, 2004

17. Samenow, J., "Weather Service Made Poor Decision in Overplaying Nor'easter Snow Predictions," The Washington Post, March 15, 2017, https://washingtonpost.com/news/capital-weather-gang/wp/2017/03/15.

18. Multivariate ENSO Index,
https://www.esrl.noaa.gov/psd/enso/mei/

19. Wintertime El Nino Pattern, NOAA Climate.gov,
https://www.climate.gov/sites/default/files/ENSO_schematicglobe_large.jpg

20. Southern Oscillation Index (SOI), National Climatic Data Center, National Oceanic and Atmospheric Administration (NOAA), June, 2018, https://www.ncdc.noaa.gov/teleconnections/enso/indicators/soi/

21. Winter (December-February) Precipitation Patterns During Strong, Moderate, and Weak El Nino Events Since 1950, https://www.climate.gov/sites/default/files/ENSO_USimpacts_prec ip.lrg.jpg

22. Burt, C. C., Extreme Weather, Chapters 1, pp. 15-43, W. W. Norton & Company, Inc., New York, 2004

Chapter 7

1. Gore, A., Inconvenient Truth: The Planetary Emergency of Global Warming and What We Can Do About It, Rodale Press, Inc. 2006

2. Ayensu, E. S., Ed., Jungles, Crown Publishers, Inc., New York, 1980, p. 7

3. Gould, S. J., Ed., The Book of Life, W. W. Norton & Company, Inc., New York, 1993, p. 44

4. Sodhi, N. S., Brook, B. W., Bradshaw, C. J. A., "Causes and Consequences of Species Extinctions," press.princeton.edu/chapters

5. Alvarez, W., T. Rex and the Crater of Doom, Princeton University Press, New York, 1997

6. Butler, R., Deforestation Rates, Mongabay, Jan. 11, 2016, https://rainforests.mongabay.com/deforestation_alpha.html.

7. Bollet, A. J., Plagues & Poxes, Demos Medical Publishing, Inc., New York, 2004, pp. 17-30

8. Alberts, B., Bray, D., Johnson, A, Lewis, J., Raff, M., Roberts, K., Walter, P., Essential Cell Biology, Garland Publishing, Inc., New York, 1998, p. 83

9. Tan, D. K. Y. and Amthor, T. S., Bioenergy, Chapter 12 in Dubinsky, Z., Ed., Photosythesis, Intech Publishing, Open Access, June 12, 2013, http://www.intechopen.com/books/photosynthesis.

10. Carbon Dioxide in Greenhouses, Ontario Ministry of Agriculture, Food, and Rural Affairs, Fact Sheet#00-077, Dec., 2002.

11. Sussman, B., Eco-tyranny, WND Books, Washington, DC, pp. 68, 248 (2012)

12. Carbon Dioxide, Centers for Disease Control and Prevention (CDC), May 1994, https://www.cdc.gov/niosh/idlh/124389.html

13. Carbon Monoxide, Centers for Disease Control and Prevention (CDC), May 1994, https://www.cdc.gov/niosh/idlh/630080.html

14. Ayensu, E. S., Ed., Jungles, Crown Publishers, Inc., New York, 1980, p. 7

15. Klein, R., Polar Ecosystem, Encyclopedia Britannica, https://www.britannica.com/science/polar-ecosystem

16. Adapted from data in: a) Scotese, C. R., Analysis of the Temperature Oscillations in Geologic Eras (2002), Scotese, C. R., 2002, http://www.scortese.com (PALEOMAP website). b) Ruddiman, W. F., Earth's Climate: Past and Future, W. H. Freeman & Sons, New York, 2001, c) Pagani, M., et al., "Marked Decline in Atmospheric Carbon Dioxide Concentrations During the Paleocene, Science, 309 (5734), 600-603, July 22, 2005.

17. Zeng, F., The Origin of Coal and World Reserves, Vol. 1 in Coal, Oil, Shale, Natural Bitumen, Heavy Oil, and Peat, Encyclopedia of Life Support Systems, www.eolss.net/Eolss-sampleAllChapter.aspx.

18. Gould, S. J., Ed., The Book of Life, W. W. Norton & Company, Inc., New York, 1993, p. 108.

19. Gould, S. J., Ed., The Book of Life, W. W. Norton & Company, Inc., New York, 1993, pp. 99-110.

20. Renne, P. R., Zhang, Z., Richards, M. A., Black, M. T., and Basu, A. R., "Synchrony and Causal Relations Between Permian-Triassic Boundary Crises and Siberian Flood Volcanism," Science, 269, pp. 1413-1416 (1995)

21. Alvarez, W., T. Rex and the Crater of Doom, Princeton University Press, New York, 1997

22. Species List: Endangered, Vulnerable, and Threatened Animals, World Wildlife Fund, www.worldwildlife.org/species/directory?sort=extinction, 5/22/17

23. Blue Whale, World Wildlife Federation, www.panda.org/what_we_do/endangered_species/cetaceans/about/blue_whale/

24. Listed Species Summary (Boxscore), U.S. Fish & Wildlife Service, ECOS Environmental Conservation Online System, https://ecos.fws.gov/ecp0/reports/box-score-report.

25. Crockford, S., Twenty Good Reasons Not to Worry About Polar Bears, The Global Warming Policy Foundation, 2015, www.thegwpf.org/content/uploads/2015/02/Crockford-2015.pdf.

26. Pennisi, E., "Polar Bear Evolution was Fast and Furious," Science, May 8, 2014, www.sciencemag.org/news/2014/05/polar-bear-evolution-was-fast-and-furious.

27. IUCC/SCC Polar Bear Specialist Group, http://pbsg.npolar.no/en/

28. Crockford, S., "Graphing Polar Bear Population Estimates Over Time," Polar Bear Science, Feb. 18, 2014,

http://polarbearscience.com/2014/02/18/graphing-polar-bear-population-estimates-over-time/.

29. Joling, D., "Walrus Won't be Listed as Threatened Species," Associated Press, Thursday, October 5, 2017.

30. Walrus Evolution, BioExpedition, Nov. 27, 2012, www.bioexpedition.com/walrus-evolution/

31. Steele, J., The 2015 Arctic Report Card: NOAA Failed Walrus Science!, Watts Up With That?, Dec. 22, 2015, https://wattsupwiththat.com/2015/12/23/the-2015-arctic-report-card-

32. Sirenko, B. I. and Gagaev, S. Y., "Unusual Abundance of Macrobenthos and Biological Invasions in the Chukchi Sea," Russ. J. Marine Bio., 33, pp. 355-364 (2007)

33. Bayer, T. and Hinerfeld, D., directors, "Acid Test: The Global Challenge of Ocean Acidification," a documentary prepared by the National Resources Defense Council (NRDC), 2009.

34. Written testimony of Sigourney Weaver, Hearing on the Environmental and Economic Impacts of Ocean Acidification before the Subcommittee on Oceans, Atmosphere, Fisheries, and Coast Guard of the United States Senate Committee on Commerce, Science, and Transportation, April 22, 2010.

35. Gould, S. J., Ed., The Book of Life, W. W. Norton & Company, Inc., New York, 1993, p. 43

36. Gould, S. J., Ed., The Book of Life, W. W. Norton & Company, Inc., New York, 1993, p. 30

37. Mahan, B. H., University Chemistry, 3rd Edition, Addison-Wesley Publishing Co., Reading, MA, 1975, pp. 218-220.

38. Bunker, B. C. and Casey, W. H., The Aqueous Chemistry of Oxides, Oxford University Press, New York, 2016, pp. 559-552.

39. Examples of useful geochemical codes include: a) Visual Minteq, Swedish Royal Institute of Technology, http://vminteq.lwr.kth.se//, b) U.S. Geological Survey's Division of Water Resources, http://water.usgs.gov/software/lists/geochemical, c) The Geochemist's Workbench Homepage, http://www.gwb.com.

40. Sharpe, S., "Aquarium Water pH Maintenance," The Spruce, https://wwww.thespruce.com/aquarium-water-ph-1378801.

41. Moore, T. G., "Global Warming: A Boon to Humans and Other Animals," Hoover Institution, https://stanford.edu/-moore/Boon_To_Man.html.

42. Burenhult, G., Ed., The First Humans, American Museum of Natural History, The Illustrated History of Humankind, Vol. 1, HarperCollins Publishers, New York, 1993.

43. Savino, J., and Jones, M. D., Supervolcano, The Career Press, Inc., Franklin Lakes, N. J. (2007), p. 132.

44. Goreham, S., "Hot Weather and Climate Change – A Mountain from a Molehill?" Heartland Institute, July 4, 2013, https://www.heartland.org/news-opinion/news/hot-weather-and-climate-change-mountain-from-a-molehill? Source=policybot.

45. Burenhult, G., Ed., Old World Civilizations, American Museum of Natural History, The Illustrated History of Humankind, Vol. 3, HarperCollins Publishers, New York, 1994.

46. Burenhult, G., Ed., Old World Civilizations, American Museum of Natural History, The Illustrated History of Humankind, Vol. 3, HarperCollins Publishers, New York, 1994. P. 137.

47. Gibbon, E., The Decline and Fall of the Roman Empire, Dell Publishing Co., Inc., New York, 1973.

48. Durant, W., The Age of Faith, The Story of Civilization, Part IV, Simon and Schuster, Inc., New York, 1950, Book IV: The Dark Ages, pp. 421-572.

49. Durant, W., The Renaissance, The Story of Civilization, Part V, Simon and Schuster, Inc., New York, 1953.

50. McWilliams, B., "The Great Fast and forgotten famine," The Irish Times, Feb. 19, 2001.

51. Bremen, L., "When Food Changed History: The French Revolution," Smithsonian.com, July 14, 2010, https://www.smithsonianmag.com/arts-culture/when-food-changed-history-the-french revolution-93598442/

52. Makyr, J., "Great Famine," Encyclopedia Britannica, https://www.britannica.com/event/Great-Famine-Irish-history.

Chapter 8

1. Prepared by Lawrence Livermore National Laboratory for the U. S. Department of Energy, LLNL, March, 2016. Data is based on DOE/EIA MER (2015). LLNO –MI-410527

2. Solar Electricity Handbook, 2017 Ed., www.solarelectricityhandbook.com/solar-irradiance.html.

3. Aggarwal, V., "What Are the Most Efficient Solar Panels on the Market?", Energy Sage, Sept. 13, 2017, https://news.energysage.com/what-are-the-most-efficient-solar-energy-panels-on-the-market/.

4. Table 6.7.B Capacity Factors for Utility Scale Generators Not Primarily Using Fossil Fuels, January 2013-October 2017, in Electrical Power Monthly, U.S. Energy Information Administration, Dec. 22, 2017, https://www.eia.gov/electricity/monthly/epm_table_grapher.php?/t =epmt_6_07_b

5. Williams, E. D., Ayres, R. U., and Heller, M., "The 1.7 Kilogram Microchip: Energy and Material Use in the Production of

Semiconductor Devices," Environ. Sci. Technol., 36, 5504-5510 (2002)

6. Linden, D., Reddy, T. B., Eds., Handbook of Batteries, Third Edition, McGraw-Hill, New York, 2002, Chapter 23.

7. Roberts, W., Photovoltaic Solar Resource of the United States, National Renewable Energy Laboratory, U.S. Department of Energy, October 20, 2008, http://wwwnrel.gov/gis/images/map_pv_national_low-res.jpg.

8. Land Requirements for Carbon-Free Technologies, Nuclear Energy Institute Policy Paper, July 9, 2015, https://www.nei.org/CorporateSite/media/filefolder/Policy/Papers/Land_Use_Carbon_Free_Technologies.pdf?ext=.pdf

9. World Bank Group, Access to Electricity (% of Population), Sustainable Energy for All Database, https://data.worldbank.org/indicator/EG.ELC.ACCS.ZS.

10. Coal, International Energy Agency, https://www.iea.org/about/faqs/coal/.

11. Hansen, M. E., Simmons, R. T., Yonk, R. M., The Unseen Costs of Solar-Generated Electricity, The Institute of Political Economy, Utah State University, April, 2016, www.usu.edu/ipe.

12. Boccard, N., "Capacity Factor of Wind Power," October 2008, https://docs.wind-watch.org/Boccard-Capacity-Factor-Of-Wind.pdf.

13. Taylor, R., "Tesla Delivers the World's Biggest Battery," Wall Street Journal, Nov. 23, 2017.

14. Produced by the National Renewable Energy Laboratory for the U. S. Department of Energy, January 9, 2012, www.nrel.gov/gis/images/80m_wind/awstwspd80onoffbigC3-3dppi600

15. Denholm, P., Hand, M., Jackson, M., Ong., S., "Land-Use Requirements of Modern Wind Power Plants in the United States," National Renewable Energy Laboratory (NREL) Technical Report NREL/TP-6A2-45834, August, 2009.

16. Levelized Cost of New Generation Resources in the Annual Energy Outlook, Energy Information Administration, July 12, 2012, http://www.eia.gov/forecasts/aeo/electricity_generation.cfm.

17. Hansen, M. E., Simmons, R. T., Yonk, R. M., The Unseen Costs of Wind-Generated Electricity, The Institute of Political Economy, Utah State University, April, 2016, www.usu.edu/ipe.

18. Fact Sheet on Altamont Pass Bird Kills, Center for Biological Diversity, https://www.biologicaldiversity.org/campaigns/...birds...altamont_pass/.../factsheet.pdf.

19. Bryce, E., "Will Wind Turbines Ever be Safe For Birds?" Audubon Society News, March 16, 2016, www.audubon.org/news/will-wind-turbines-ever-be-safe-for-birds.

20. Summary of Wind Turbine Accident Data to 30 September 2017, Caithness Windfarm Information Forum, Sept. 30, 2017, www.caithnesswindfarms.co.uk/accidents.pdf.

21. Ethanol Fuel from Corn Faulted as 'Unsustainable Subsidized Food Burning' in Analysis by Cornell Scientist, https://news.cornell.edu/stories/2001/08/ethanol-corn-faulted-energy-wasted-scientist-says.

22. Pimentel, D., Patzek, T. W., "Ethanol Production Using Corn, Switchgrass, and Wood; Biiodiesel Production Using Soybean and Sunflower," Natural Resources Research, 14, pp. 65-76, March, 2005.

23. Weast, R. C., Ed., CRC Handbook of Chemistry and Physics, 58th Ed., CRC Press Inc., Cleveland, Ohio, 1977, pp. D274-279.

24. Greenhouse Gas Emissions Estimation Methodologies for Biogenic Emissions from Selected Source Categories: Solid Waste Dispposal, Wastewater Treatment, Ethanol Fermentation, Draft Report to the U.S. EPA, RTI International, Dec. 14, 2010, www3.epa.gov/ttnchiel/efpac/ghg/GHG_Biogenic_Report_draft_Dec1410.pdf.

25. Economides, M., "Corn-based Ethanol: The Real Cost," Fuel Fix, May 13, 2011, http://fuelfix.com/blog/2011/05/13/corn-based-ethanol-the-real-cost/

26. Fill Up On Facts: Renewable Fuel Standards, http://www.filluponfacts.com/#/?section=outdoor-power-equipment-and-the-rfs.

27. Alexander, D., "U.S. Air Force Tests Biofuel at $59 per Gallon," Reuters, July 14, 2012, www.reuters.com/article/us-usa-minitary-biofuels/u-s-air-force-tests-biofuel-at-59-per-gallon-idUSBRE86E01N20120715

28. Johnson, T., "New E15 Gasoline May Damage Vehicles and Cause Consumer Confusion," AAA NewsRoom, Nov. 30, 2012, http://newsroom.aaa.com/2012/11/new-e15-gasoline-may-damage-vehicles-and-cause-consumer-confusion.

29. Clark, M., "Aging U.S. Power Grid Blacks Out More Than Any Other Developed Nation," International Business Times, July, 17, 2014, http://www.ibtimes.com/aging-us-power-grid-blacks-out-more-any-other-developed-nation-1631086.

30. Population Density of United States Dot Map.jpg, Environmental Science Blog, Thurs., Sept. 22, 2011, www.enb150-2011f-jf.blogspot.com/2011_09_01archive.html.

31. Orr, I., Wall Street Journal, July 18, 2017, p. A13

32. Alvarez, G. C., Jara, R. M., Julian, J. R. R., and Bielsa, J. I. G., Executive Summary: Lessons from the Spanish Renewables Bubble, Study on the Effects on Employment of Public Aid to

Renewable Energy Sources, Universidad Rey Juan Carlos, March 2009.

33. CO_2 Emissiions (kt), The World Bank, https://data.worldbank.org/indicator/EN.ATM.CO2E.KT?view = map.

34. Rochelle, G. T., "Amine Scrubbing for CO_2 Capture," Science, 325, pp. 1652-1654, Sept. 25, 2009.

35. Lomborg, B., "Climate-Change Policies Can be Punishing for the Poor," Wall Street Journal, Jan. 5, 2018, p. A15.

36. Schow, A., "President Obama's Taxpayer-Backed Green Energy Failures," Daily Signal, October 18, 2012, http://dailysignal.com/2012/10/18/president-obamas-taspayer-backed-green-energy-failures/

37. U.S. Coal Production, 2011-2017, Quarterly Coal Report, April-June-2017, U.S. Energy Information Administration, https://www.eia.gov/coal/production/quarterly/pdf/t1p01p1.pdf.

38. Walters, D., "Cap-and-Trade Feeds Funding Frenzy Among Politicians," The Mercury News, Aug. 30, 2017, https://mercurynews.com/2017/03/30/walters-cap-and-trade-feeds-funding-frenzy-in-sacramento.

39. Sexton, R. and Sexton, S., "The Fatal Flaw in California's Cap-and-Trade Program," Wall Street Journal, Oct. 20, 2017.

40. Baker, D. R., "California's Greenhouse Gas Emissions Fall by Less Than 1%," San Francisco Chronicle, June 9, 2017, http://www.sfgate.com/business/article/California-s-greenhouse-gas-emissions-fall-by-11206585.php.

41. "California Prays to the Sun God," Wall Street Journal, May 12-13, 2018, p. A12

42. New U.S. Urban Area Data Released, October 24, 2017, www.newgeography.com/content/002747-new-us-urban-area-data-released.

43. Mahan, B. H., University Chemistry, 3rd Ed., Addison-Wesley Publishing Company, Reading, Mass., 1975, p. 846

44. Safety of Nuclear Power Reactors, World Nuclear Association, May 2016, www.world-nuclear.org/information-library/safety-and-security/safety-of-plants/safety-of-nuclear-power-reactors.aspx.

45. Radioactive Waste Management, World Nuclear Association, June 2017, www.world-nuclear.org/information-library/nuclear-fuel-cycle/nuclear-wastes/radioactive -waste-management.aspx.

46. "Uranium 2014: Resources, Production, and Demand," A joint report issued by the OEDC Nuclear Energy Agency (NEA) and the International Atomic Energy Agency (IAEA).

47. Ferguson, W., "Record Haul of Uranium Harvested from Seawater," New Scientist, Issue 2880, Aug. 29, 2012.

48. Thorium, World Nuclear Association, February 2017, www.world-nuclear.org/information-library/current-and-future-generation/thorium.aspx.

49. Mahan, B. H., University Chemistry, 3rd Ed., Addison-Wesley Publishing Company, Reading, Mass., 1975, p. 846

50. Weast, R. C., Ed., CRC Handbook of Chemistry and Physics, 58th Ed., CRC Press Inc., Cleveland, Ohio, 1977, p. B271

51. www.lockheedmartin.com/us/products/compact-fusion.html

Chapter 9

1. Montford, A., Fraud, Bias, and Public Relations: The 97% 'Consensus' and Its Critics, The Global Warming Policy Foundation, 2014, www.thegwpf.org.

2. DeWeese, T., The Heidelberg Appeal, American Policy Center, March 29, 2002, https://americanpolicy.org/2002/03/29/the-heidelberg-appeal/

3. Global Warming Petition Project, www.petitionproject.org/

4. Revelle, R., and Suess, H., "Carbon Dioxide Exchange Between Atmosphere and Ocean and the Question of an Increase in Atmospheric CO_2 During the Past Decades," Tellus, 9, 18-27 (1957).

5. Singer, S. F., Revelle, R., and Starr, C., "What To Do About Greenhouse Warming: Look Before You Leap," Cosmos: A Journal of Emerging Issues, 5, Summer 1992.

6. Sussman, B., Climategate, WND Books, Washington, D.C., 2010, pp. 90-93.

7. John Coleman Debunks the Myth of Global Warming, YouTube, Dec. 26, 2009, https://www.youtube.com/watch?v=D8FhmuWWcGw

8. Steyn, M., Ed., "A Disgrace to the Profession", Vol. 1, Stockade Books, Woodsville, New Hampshire, 2015, p. 22

9. Steyn, M., Ed., "A Disgrace to the Profession", Vol. 1, Stockade Books, Woodsville, New Hampshire, 2015, p. 146

10. Steyn, M., Ed., "A Disgrace to the Profession", Vol. 1, Stockade Books, Woodsville, New Hampshire, 2015, pp. 31-32

11. Steyn, M., Ed., "A Disgrace to the Profession", Vol. 1, Stockade Books, Woodsville, New Hampshire, 2015, pp. 137-138

12. Steyn, M., Ed., "A Disgrace to the Profession", Vol. 1, Stockade Books, Woodsville, New Hampshire, 2015, p. 86

13. "A Factual Look at the Relationship Between Climate and Weather," Subcommittee on Science, Space, and Technology, Dec.

11, 2013, Testimony of John R. Christy, University of Alabama in Huntsville, https://www.nsstc.uah.edu/users/john.christy/docs/ChristyJR_Written_131211_01.pdf.

14. Lamb, H., "The Early Medieval Warm Period and Its Sequel," Paleogeography, Paleoclimatology, Paleoecology, 1, pp. 13-37 (1965)

15. Lamb, H. H., Climate, History, and the Modern World, 2nd Ed., Taylor & Francis, 1995.

16. International Panel on Climate Change Third Assessment Report, 2001, p. 29, Fig. 2.3.

17. Mann, M. E., Bradley, R. S., Hughes, M. K., "Global-Scale Temperature Patterns and Climate Forcing Over the Past Six Centuries," Nature, 392, 779-787 (1998).

18. John R. Christy, University of Alabama in Huntsville, "Examining the Process Concerning Climate Change Assessments," testimony before the House Science, Space and Technology Committee, March 31, 2011, https://science.house.gov/sites/republicans.science.house.gov/files/documents/hearings/ChristyJR_written_110331_all.pdf.

19. a) Monckton, C., Climategate: Caught Green-Handed, Science and Public Policy Institute Original Paper, scienceandpublicpolicy.org, Dec. 7, 2009; b) Sussman, B., Climategate, WND Books, WorldNetDaily, New York, 2010.

20. Mann, M. E., Bradley, R. S., Hughes, M. K., "Global-Scale Temperature Patterns and Climate Forcing Over the Past Six Centuries," Nature, 392, 779-787 (1998)

21. International Panel on Climate Change Third Assessment Report, 2001, p. 29, Fig. 2.3

22. Steyn, M., Ed., "A Disgrace to the Profession", Vol. 1, pp. iii-vi, Stockade Books, Woodsville, New Hampshire, 2015

23. Richard, T., "How NOAA Rewrote Climate Data to Hide Global Warming Pause," Frontiers of Freedom, Feb. 23, 2016, httpe://www.ff.org/how-noaa-rewrote-climate-data-to-hide-global-warming-pause/.

24. Moran, R., "Judicial Watch Files Suit Over NOAA Climate Docs," American Thinker, Dec. 22, 2015, http://www.americanthinker.com/blog/2015/12/judicial_watch_files_suit_over_noaa_climate_docs.html.

25. Climategate Update: Judicial Watch Sues for Records between Key Obama Administration Scientists Involved in Global Warming Controversies, Judicial Watch, March 27, 2017, https://www.judicial watch.org/press-room/press-releases/climategate-u...bama-administration-scientists-involved-global-warming-controversies/.

26. Heller, T., "The History of NASA/NOAA Temperature Corruption," posted at https://realclimatescience.com/2016/01/the-history-of-nasanoaa-temperature-corruption/

27. U.S. Committee on Science, Space, and Technology, "Former NOAA Scientist Confirms Colleagues Manipulated Climate Records," Press Release, Feb. 5, 2017, https://science.house.gov/news/press-releases/former-noaa-scientist-confirms-colleagues-manipulated-climate-records.

28. Heller, A., "NOAA Data Tampering Approaching 2.5 Degrees," Real Climate Science, March 20, 2018, https://realclimatescience.com/2018/03/noaa-data-tampering-approaching-2-5-degrees/

29. Brakey, M., "Black Swan Climate Theory," in Gosselin, P., "151 Degrees of Fudging. . . Energy Physicist Unveils NOAA's "Massive Rewrite" of Maine Climate History," NoTricksZone,

May 2, 2015, http://notrickszone.com/2015/05/02/151-degrees-of-fudging-energy-physicist-unveils-noaas-massive-rewrite-of-maine-climate-history/#sthash.UvyUKRCC.dpbs.

30. Van Biezen, M., The Most Comprehensive Assault on 'Global Warming' Ever, Daily Wire, Dec. 23, 2015, http://www.dailywire.com/news/2071/most-comprehensive-assault-global-warming-ever/ pp. 6-12.

31. Monckton, C., Climategate: Caught Green-Handed, Science and Public Policy Institute Original Paper, scienceandpublicpolicy.org, Dec. 7, 2009, pp. 32-35.

32. Claim That Sea Level Is Rising Is a Total Fraud, Interview: Dr. Nils-Axel Morner, Executive Intelligence Review, June 22, 2007, pp. 33-37, www.larouchepub.com/eiw/public/2007/eirv34n25-20070622/33-37_725.pdf

33. Watts, A., "No, Climate Change Didn't Cause "5 Whole Pacific Islands" to be Swallowed by Sea Level Rise,' March 8, 2017, https://wattsupwiththat.com/2017/03/08/no-climate-change-didn't-casue-5-whole-pacific-islands-to-be-swallowed-by-sea-level-rise.

34. Albert, S., Grinham, A., Gibbes, B., Leon, J., Church, J., "Sea Level Rise Swallows 5 Whole Pacific Islands,", The Conversation, Scientific American, May 9, 2016.

35. Goddard, S., "NASA Sea Level Fraud," Real Science, https://stevengoddard.wordpress.com/nasa-sea-level-fraud/

36. Heller, T., "NASA – Doubling Sea Level Rise by Data Tampering," April 25, 2016, https://realclimatescience.com/2016/04/nana-doubling-sea-level-rise-by-data-tampering/

37. Heller, T., "NASA Shows How Science Fraud is Done," Oct. 25, 2016, https://climatism.blog/2016/10/27/nasa-sea-level-rise-fraud/

38. TV documentary *Doomsday Called Off*, Director: Lars Mortensen, Denmark (2004)

39. Interview: Dr. Nils-Axel Morner, "Sea-level Expert: It's Not Rising!," 21st Century Science & Technology, Fall 2007, pp. 25-29, http://www.21stcenturysciencetech.com/articles%202007/MornerInterview/

40. Hughes, W., "Did Australian scientists/students pull down this tree in the Maldives?" IPCC, News and Views, Sea Levels, June 25, 2007, warwickhughes.com. (This reference is not particularly conclusive.)

41. Watts, A., "Despite popular opinion and calls to action, the Maldives are not being overrun by sea level rise," March 19, 2009, wattsupwiththat.com.

42. Spry, J., "Shock News: Satellite Showed Little Sea Level Rise Before They Tampered with the Data," Climatism, Nov. 14, 2013, https://climatism.blog/2013/11/14/shock-news-satellite-showed-little-sea-level-rise-before-they-tampered-with-the-data/

43. Goddard, S., "Falling Sea Level," Real Science, March 23, 2016, https://stevengoddard.wordpress.com/2016/03/23/falling-sea-level/

44. Durden, T., "NASA Confirms Falling Sea Levels for Two Years Amidst Media Blackout," Zerohedge, July 27, 2017, https://www.zerohedge.com/news/2017-07-27/nasa-confirms-falling-sea-levels-two-years-amidst-media-blackout/

45. Peltier, W. R., "Closure of the Budget of Global Sea Level Rise Over the GRACE Era: The Importance and Magnitudes of the Required Corrections for Global Isostatic Adjustment," Quart. Sci. Rev., 28 (17-18), 1658-1674 (2009)

46. Reager, J. T.,

INDEX

over geologic time, 46-47
over past 10,000 years, 47
primeval, 46
consumed by organisms, 59
contained in geologic
formations, 60-62
dissolved in ocean water, 52
exhaled by humans, 58
exhaled by organisms, 58
factors controlling
atmospheric levels of, 50-65
  biological factors, 58-60
  human activities, 51, 65
  oceanic equilibria, 52-55
  volcanic activity, 62
from fossil fuel emissions, 4,
5, 51
impact on climate
  actual, 49-50
  predicted, 48
impact on oceanic acidity,
127-130
in photosynthesis, 13, 114
need for sustaining life, 13,
114-116
on Venus, 48-49
solubility in water, 52-55,
209
toxicity to humans, 116-117
worldwide emissions of, 5
Carbonate minerals, 208
  CO_2 contained in, 61
  precipitation of, 213-214
Carbonated beverages, CO_2
dissolved in, 53
Carbonic acid, 127, 208, 211-
212
Carboniferous Period, 119,
120

Center for Disease Control,
116
Christie, John, on global
warming, 172, 173
Climatic temperature
variations
  impact on humanity, 131-
  134
  over millions of years, 29-
  31, 47
  over 10,000 to 100,000
  years, 31-34
  over less than 10,000 years
  34-41
  since 1880, 41, 96-97
  versus carbon dioxide, 48-50
Climategate, 10-11, 174-177
Cover-ups of scientific fraud,
175-176
Falsification of climate
results, 174-175
Intimidation of scientists,
176-177
Perversion of peer review,
176
Climatic Research Unit
(CRU), 173, 174-177
Clouds
  as absorbers of infrared
  radiation, 27
  impact on Earth's energy
  cycle, 21
Coal, deposits and worldwide
inventory of, 51, 119
Coleman, John, on global
warming, 171
Continental drift, 29-31
Cook survey, on global
warming consensus, 169-170